Prehistoric Peoples of South Florida

Prehistoric Peoples of South Florida

William E. McGoun

The University of Alabama Press
Tuscaloosa and London

Copyright © 1993
The University of Alabama Press
Tuscaloosa, Alabama 35487-0380
All rights reserved
Manufactured in the United States of America

designed by zig zeigler

∞

The paper on which this book is printed meets the
minimum requirements of American National Standard for
Information Science-Permanence of Paper for
Printed Library Materials,
ANSI Z39.48-1984.

Library of Congress Cataloguing-in-Publication Data

McGoun, William E., 1937–
Prehistoric peoples of South Florida / William E. McGoun.
p.   cm.
Includes bibliographical references and index.
ISBN 0-8173-0686-2 (alk. paper)
1. Indians of North America—Florida—Antiquities.
2. Florida—Antiquities.   I. Title.
E78.F6M39   1993
975.9'01—dc20   92-40833

2 4 6 5 3
01 03 05 04 02

British Library Cataloguing-in-Publication Data available

Contents

Illustrations

The Theory and the Area

To many people in South Florida, an oldtimer is someone who has been around for more than five years, and nothing worth talking about happened before the dawn of the twentieth century. In the following I will seek to set the record straight, to recount for oldtimer and newcomer alike the millennia of prehistory that are just now beginning to emerge from the mists.

To the average South Floridian, Indian means Seminole. The Seminoles in fact did not arrive in Florida until nearly two centuries after the Spanish. The cultures of which I will tell became extinct when Spain ceded Florida to Great Britain in 1763, just half a century after the first appearance of Seminoles. Christopher Columbus is a rank newcomer compared to the first Floridians.

In reaching that point, I will tell of the big-game hunters who arrived in South Florida at least 10,000 years ago, the bands of hunters and gatherers who subsisted in harmony with their environment for thousands of years, the horticulturists who built earthworks near Lake Okeechobee about the time of Christ, the coastal gatherers who exploited the shellfish beds of Charlotte Harbor area to build complex societies and structures, and finally the groups of the early historic period, including the mighty Calusa who killed the first known European voyager to Florida and held off intruders until finally overwhelmed in the eighteenth century.

This work is designed to be at the same time a scholarly dissertation for the specialist and a book that will be of interest to the lay public. I believe that this is a good time to try to reach the layperson with this rich story because of the growing interest in preservation of cultural resources. Perhaps this work will increase the awareness of South Florida's prehistoric heritage and strengthen public support for preserving as much of that heritage as possible.

To that end I have placed footnotes for discussion of some of the more arcane matters referred to within the chapter. These notes allow the reader to check the sources of my information and to become more familiar with some of the differing points of view regarding the prehistory of South Florida.

It may seem to some readers that the text becomes bogged down at points in discussions of projectile points, and especially of pottery. The problem is that points and pottery are the two great guideposts of the prehistorian, who by definition works in the absence of written records. Each is durable and each is amenable to variations in style that suggest change over time and space. These changes often are the only means by which the evolution, expansion, and contraction of cultures can be detected.

Behind these points and pottery sherds are people, people who organized themselves first into small bands of no more than a score who wandered together in search of food, then into larger egalitarian villages exploiting the sea or the land effectively enough to stay put at least most of the time, and finally into powerful and complex groups complete with kings and nobles and priests.

In analyzing human cultural development, I believe a culture can be understood fully only in terms of its own history, a theory known as cultural historical. That is not to say cultures change for arbitrary reasons. Changes arise and are accepted because they help people adapt or adjust to their environment, though it would be wrong to assume baldly that environment determines culture. For one thing, given environmental conditions can lead to more than one type of response. A coastal culture confronted by rising sea level can either put its homes on stilts or move landward, for instance. In many cases, the precise response cannot be explained fully without knowledge of the culture's history.

Ideas persist long after the conditions that give rise to them have changed. Frequently, but not always, this phenomenon occurs for religious or ideological reasons. Ideology often is a useful tool for anthropologists in that it helps trace the origins of cultures. To use a present-day secular example, if archaeologists 10,000 years hence had no idea of the origin of most twentieth-century South Floridians, they could deduce a migration from the north from the way those ancients shunned plants suited to their local climate in favor of exotic northern plants that survive and thrive only with more care. It is because of our history that we insist on using as much or more water on our lawns as we do inside our houses, even though we contend with repeated droughts.

Also, it is important to note that traits are accepted because the people believe they are adaptive, and the people sometimes are wrong. The Ghost Dance offered hope to nineteenth-century Plains Indians who saw no other way to keep their culture alive in the face of white intrusion, but at least one group soon realized to its sorrow that the special shirts its members wore were not impervious to cavalry bullets.

Finally, cultures confronted with environmental change—and here I use *environment* in a broad sense to include both the natural environment and other human societies—have a third choice beyond moving and adapting: They can make the environment adapt to them. While the trend in this direction has become accelerated with the development of modern machinery and technology, it is by no means a recent development. The big-game hunters of 10,000 years ago may have set fires in order to clear off old growth and allow the new green shoots that game animals like to eat. At least one reason for many of the earthworks in the Lake Okeechobee Basin was to lift agricultural fields above the reach of ground-water.

The explanations for cultural traits are to be found ultimately in environmental conditions, but the connections are frequently long and convoluted and involve what people believe to be true as well as what is true. Only a full knowledge of a culture's history can provide a full explanation, and this is not possible given the limits of the archaeological record. Much is and will remain conjectural.

In speaking of South Florida, the first task is to define it. The Florida Legislature, for instance, seems to include Tampa, inasmuch as it established the University of South Florida there. To the average Miamian, however, anything north of West Palm Beach is excluded.

In my definition I rely heavily upon the distribution of the distinctive aboriginal sand-tempered gritty pottery ware known as the Glades Series. I will strive, however, to restrict my use of the word *Glades* to the Everglades region of the interior. I think it is confusing to apply geographical or especially cultural names loosely, a problem demonstrated most graphically by the use of the name *Calusa* for pottery that is not even found in the heartland of the historic Calusa.

Thus defined, South Florida at its greatest extent was that area south of a line running roughly southwestward from Cocoa to Bradenton. Except around Lake Okeechobee, this boundary corresponds roughly to the southern limit of winter freezes. That fact may have some cultural significance though at the moment any correlation is purely speculative. There can be no neat boundary—prehistoric Floridians did not share our concern with property lines—and the border certainly fluctuated over time. The affiliations of the Peace River basin remain largely speculative, and the Manatee-Sarasota region is definitely one of differing influences over time.

Figure 1 delineates roughly the three regions into which I have divided South Florida: Caloosahatchee, Lake Okeechobee, and Southeast Florida. What is immediately notable is that most of South Florida falls into none of these areas. Little of the interior away from Lake Okeechobee ever was settled. Other coastal areas were subject to different influences at different times. The area south of Caloosahatchee, which includes the famous Key Marco site, shows affinities to Southeast Florida until A.D. 800 and to the Caloosahatchee region afterward. The area north of Caloosahatchee becomes less similar to South Florida over time, as does the area north of Southeast Florida on the east coast (though there is some fluctuation in this case). Affiliations of the Kissimmee River area (north of Okeechobee Basin) are somewhat unclear though it seems to have been tied to the Okeechobee Basin after A.D. 500 or later.

The Cocoa-Bradenton line seems to work best in historic

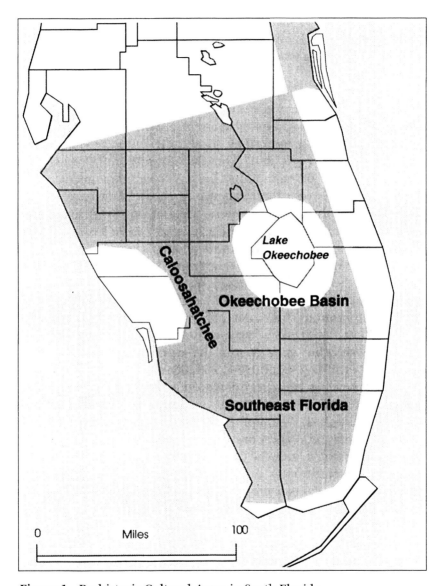

Figure 1. Prehistoric Cultural Areas in South Florida

times. It has the virtue of corresponding reasonably well to the limits of Calusa influence and also serves to divide those historic groups to the north who had horticulture from those to the south who did not.

Over the years there have been various other attempts to organize prehistoric South Florida in terms of cultural areas. These will be discussed in subsequent chapters.

As for prehistoric periods in South Florida, there is a general framework of Paleoindian, Archaic, Transitional, and Glades with a number of variations, most minor, depending upon whose system is considered. Some of those variations will be discussed in subsequent chapters. Most of our knowledge concerns the Glades period because it is the most recent and is marked by the presence of ceramics. The various Glades periods first were defined by John M. Goggin in 1947. He set no starting date for Glades I, but terminated it between A.D. 900 and 1000 on the basis of the appearance of incised ceramic decoration and the development of shell tools. He set the advent of Glades III at A.D. 1200 on the bases of greater cultural development, more contact with other groups, and the near-disappearance of incised decorations (Goggin 1964b:80, 86–87).

The system has been reconsidered at various times over the years, with two of those reconsiderations being published in 1988. Randolph J. Widmer (1988:80–81) puts the advent of Glades I at 500 B.C. based on the presence of villages, an abundance of ceramics, and the peopling of all of South Florida. He uses the same criterion as Goggin for Glades II but places the advent at A.D. 700 based on more recent evidence. He follows Goggin in dating Glades III from A.D. 1200 for essentially the same reasons. John W. Griffin (1988:128) agrees, except that he would not begin Glades I until A.D. 1.

I prefer Widmer's date of 500 B.C. for the beginning of Glades I, due to the appearance of plain sand-tempered pottery about that time. Where I differ with Widmer is in his choice of an A.D. 700 advent for Glades II based upon the appearance of some incised pottery types (Widmer 1988:Table 2). I view the disappearance about a century later of many incised types, probably due to political expansion of the Charlotte Harbor people (a

Table 1. Cultural Periods in Prehistoric South Florida

Period	Dates	Characteristics
Paleoindian	10,000–7000 B.C.	distinctive lithic artifacts; evidence of human presence in association with remains of extinct Pleistocene animals
Archaic		
Early	7000–5000 B.C.	stemmed projectile points
Middle	5000–3000 B.C.	sites in Southeast Florida
Late	3000–1500 B.C.	more sites, notably in Southeast Florida
Transitional	1500–500 B.C.	fiber-tempered pottery appears
Glades I	500 B.C.–A.D. 800	sand-tempered pottery; rim incising
Glades II	A.D. 800–1200	incising of rims dies out on Gulf Coast, persists on East Coast
Glades III	A.D. 1200–1566	incising decreases; rim-tooling appears; check-stamping appears
Historic	A.D. 1566–1763	European objects

point that will be discussed in Chapter 5), to be more significant. Therefore I would date the beginning of Glades II at A.D. 800 (see Table 1).

All prehistoric human development in South Florida can be considered of a single tradition, in that a tradition is defined as a long-lasting manifestation of certain core cultural features; Willey and Phillips (1958:37) refer to "persistent configurations in single technologies or other systems of related forms." In this case the core is simply adaptation to the South Florida environment. This tradition is divided into periods, defined as shorter spans of time in which more common cultural features can be noted than in a tradition. Periods in turn are divided into stages, which are again shorter in time and more specific in detail. There is no easy way to say what constitutes a period or a stage;

7

the divisions will always be to some extent arbitrary. Finally, the area in which a tradition can be detected at any given time is its horizon.

A culture is a group of people, a society, along with those materials it uses and the ideas under which it operates. In popular usage the terms *culture* and *society* often are interchanged but to the specialist they must be kept distinct.

ADDENDUM TO THE INTRODUCTION

The most spectacular new find since the first printing of this book in 1993 is the Miami Circle, a trench 38 feet in diameter carved in limestone on the south bank of Miami River, across from the site of the 16th century village of Tequesta. Most archaeologists believe it is aboriginal, possibly dating to the time of Christ, though Jerald T. Milanich ("Much Ado About a Circle," *Archaeology* 52[5]: 22–25, 1999) suggests it could be associated with a 1950s septic-tank drain field.

Additional canals have been reported by Ryan J. Wheeler ("Aboriginal Canoe Canals of Cape Sable," *Florida Anthropologist* 51[1]: 15–24, 1998) and George M. Luer ("The Naples Canal: A Deep Indian Canoe Canal in Southwestern Florida," *Florida Anthropologist* 51[1]: 25–36, 1998).

Julian Granberry ("The Position of the Calusa Language in Florida Prehistory: A Hypothesis," *Florida Anthropologist* 48[3]: 156–173, 1995) believes all South Florida aborigines of the 16th century spoke dialects of a single language and that that language may have been related to the Tunica language of the lower Mississippi River valley. The idea is intriguing, but the evidence—no more than 57 words—is thin.

Robert Patton ("Mississippian and Calusa Political Development," paper presented at the 56th annual meeting of the Southeastern Archaeological Conference, Pensacola, Florida, Nov. 10–13, 1999) suggests a third possible explanation for Calusa political development, influence from the complex late prehistoric societies of the Mississippi Valley. Randolph J. Widmer believes the presence of rich but circumscribed seafood resources necessitated strong leadership (p. 35, herein) whereas William H. Marquardt has suggested that complexity may have arisen as a defense against the Spanish (p. 16n, herein).

The only significant change I would make in the text were I writing the book today is to say flatly that the Archaic people of South Florida were newcomers and not descendants of the Paleo-Indians, a question I left open (pp. 50–51, herein) in the first printing. To me the decisive factor is the difference in head shapes as first demonstrated by T. Dale Stewart (p. 41, herein). I do not, however, believe that Paleo-Indians came from Europe, as do some archaeologists quoted by Sharon Begley and Andrew Murr ("The First Americans," *Newsweek*, pp. 50–57, April 26, 1999).

ONE

Caciques and Conquistadors

Aboriginal Peoples in the Menéndez Period

Neither the cacique nor the conquistador was taking any chances. The two Spanish brigantines were drawn up broadside to the shore with their artillery concentrated on the landward side and a large supply of hail-shot on hand. When the conquistador disembarked he was accompanied by thirty soldiers, each with his fuse lit so he could fire his matchlock arquebus quickly. Against this superior firepower the cacique arrayed superior manpower. He was accompanied by 300 archers as he took his seat on the platform the conquistador had provided on the beach.

Thus did Gonzalo Solís de Merás (1964:141) describe the meeting at which prehistory gave way to history in South Florida. The conquistador was Solís de Merás's brother-in-law, Pedro Menéndez de Avilés. As adelantado of Florida, Menéndez had a franchise to explore Florida and profit from its resources as long as he followed certain Spanish guidelines.

Five months previously, in September of 1565, Menéndez had established at St. Augustine what would become the first permanent European settlement in today's mainland United States of America (Solís de Merás 1964:88–89), subsequently routing the French who had established a foothold on the St. Johns River (Solís de Merás 1964:115–127).

Menéndez had more than one reason for visiting the cacique (a term the Spanish borrowed from the Taino culture of Hispaniola) at his headquarters on the Bay of Juan Ponce in Southwest Florida. (Most believe that is today's Estero Bay and that the cacique's capital was on Mound Key, though some hold out for Charlotte Harbor and the Pine Island area.)[1] One very

---

[1]Anthropologists initially had assumed that the Calusa capital was within Charlotte Harbor, presumably because that harbor commands the approach to the

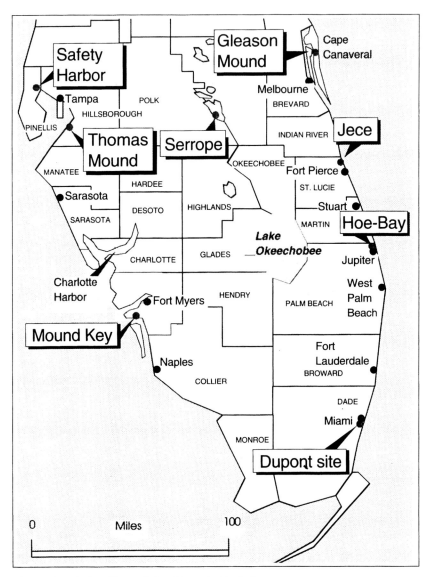

Figure 2. Major Sites of the Historic Period

personal reason was to find his son, Don Juan Menéndez, whose ship had foundered possibly on this very coast (Solís de Merás 1964:68).

Don Juan Menéndez was not the only Spaniard who might be found. The cacique's sway extended southward to the Florida Keys and northward at least on occasion to the Tampa Bay area, putting him close to the Havana-Veracruz shipping lanes (Wright 1981:40). His people had captured possibly 250 ship-wrecked Spaniards over the years (Connor 1925:39) and offered up a number of them as human sacrifices (Solís de Merás 1964:139). Escalante Fontaneda, another Spaniard who had been held captive, told of the killing of forty-two captives, including his older brother (Menéndez de Aviles, as translated in Hann 1991:301).

The younger Menéndez would not be found, even though Fontaneda told of having seen him (Fontaneda 1973:33). Solís de Merás (1964:139) says the rescue of captives was the primary reason for the conquistador's visit to the cacique. Later he would have another reason. In the summer of 1566, during a trip up the St. Johns River, he would be told of a cross-state waterway that emptied into the Gulf of Mexico in the cacique's territory (Solís de Merás 1964:205). This made a Spanish presence important for strategic reasons vis-à-vis the French (Lyon 1976:141–142).

The cacique was Carlos II, leader of the Calusa, and he had his own reasons for wanting to talk. The Calusa's chief rivals for power were the Tocobaga, whose headquarters were at what is now known as Safety Harbor on Old Tampa Bay (Lewis 1978:54). Carlos saw in these well-armed visitors a valuable ally in smash-

---

Caloosahatchee River. In the 1940s, Goggin (m.s.:48) had placed the capital "near the mouth of the Caloosahatchee River, possibly in Pine Island Sound." By 1964 he had changed his mind; both he and Sturtevant (1964:182) and later Clifford M. Lewis (1978:19) opted for Mound Key. They rely upon the account of López de Velasco (translated in Hann 1991:311–312), who describes a bay fourteen to eighteen miles in circumference and an island two miles in circumference, descriptions that fit Estero Bay and Mound Key but neither Charlotte Harbor nor any island within it.

ing the Tocobaga once and for all (Solís de Merás 1964:223). That the rivalry was longstanding is shown by the account of Juan Ortiz, a Spaniard who was held captive by the Tocobaga people to the north for twelve years and who tells of a Calusa raid and Tocobaga counter-raid prior to 1539 (Steele 1972:120, 123).[2] Archaeological evidence suggests a Calusa incursion into the Tampa Bay area in historic times (Willey 1982:120, 124), and Spanish sources tell of back and forth struggles during the 1560s (Lewis 1978:26).

Carlos also may have wanted an ally against rivals within the Calusa. His father, Carlos I, had usurped the chieftainship (Goggin and Sturtevant 1964:123), and Carlos himself was viewed by some as an usurper (Lewis 1978:33). Spanish accounts tell of plottings, revolts, and assassinations (Zubillaga 1946:306–308, 310–311, 337, as translated in Goggin and Sturtevant 1964:194), Solís de Merás says Carlos' subjects feared Don Felipe, the Calusa "captain general," more than they did the cacique himself (Solís de Merás 1964:151), and Stephen Edward Reilly believes the Carlos-Felipe rivalry was the chief reason the cacique sought the alliance (Reilly 1981:409–410). Later, Felipe would tell the Spanish that he was the designated heir of Carlos

[2]The extent of Calusa control northward on the Gulf Coast is a matter of conjecture, given the lack of evidence for historic occupation between Charlotte Harbor and Tampa Bay, a point noted by both Goggin and Fairbanks in coming to different conclusions. Goggin (m.s.:48) believed the Tocobaga capital to be on the north shore of Charlotte Harbor, while Fairbanks (1957:42) placed the Calusa's northward limit at the south end of Tampa Bay.

It should be noted that the basis on which I accept the Fairbanks view, discovery of two of the distinctive metal tablets in the south end of Hillsborough County (Willey 1982:124), is not universally accepted. I have said (1981:33) the presence of such tablets indicates Calusa control, even if only of a transient nature. Others have argued that the tablets were found in areas the Calusa did not control and that they probably were religious rather than political in nature (Allerton et al. 1984:15). I would counter that the lack of Calusa control is a conjecture contradicted by both historical accounts and the presence of the tablets and that the sacred-secular distinction appears from Spanish records and was meaningless to the Calusa.

I's predecessor and that thus both Carloses had assumed an office that was rightfully his (Zubillaga 1946:309–311, as translated in Goggin and Sturtevant 1964:193).

Nomenclature is a source of confusion. In Lewis's system, "*Calusa* is used to designate the tribe or culture, *Calos*, the principal town, and *Carlos*, the name of the principal chief" (Lewis 1978:19; emphasis is his). I follow that system regarding the culture and its chief, while referring to the town as simply "the Calusa capital" to lessen confusion. As for the Tocobaga, "Often, the same name is used for the province, the largest town and the chief" (Bullen 1978:50). The only way out is to refer simply to "the Tocobaga people" or "the Tocobaga leader" while calling the capital by its present-day name, Safety Harbor.

Neither the cacique nor the conquistador got what he wanted. Menéndez did take Carlos to Safety Harbor on his ship but refused to engage in military operations against the northerners (Solís de Merás 1964:223, 228). As for the internal rivalry, both Carlos and Felipe would wind up dead by Spanish hands within three years (Zubillaga 1946:296, as translated in Lewis 1978:29; Connor 1939:39). Further, the rivalry never worked out to the Spaniards' advantage. In a showdown, the Calusa would put aside whatever differences they had internally to present a united front against the Spanish; "'Divide and conquer' was not an alternative (for the Spanish) in South Florida," says Jeremy Stahl (1986:174).

Carlos did agree to free the captives he held, but only after Menéndez had tricked him into coming aboard ship where the Spanish had the upper hand. Carlos did enter into an alliance with the Spanish sealed by the gift of his sister (Solís de Merás 1964:142, 144). But, neither the marriage nor the alliance endured.

As a Catholic, Menéndez was reluctant to consummate a polygynous union with Doña Antonio, though he may have done so; Solís de Merás is ambiguous on this point (1964:149–151, 190). While the Calusa were receptive to outside ideas of which they approved (Lewis 1978:44), they were fiercely independent and would not countenance the constant interference

13

in their affairs by the military-clerical settlement the Spanish established among them.

The Calusa leaders were willing to add the Spanish god to their pantheon, but not to give up their traditional deities, as the Jesuit Father Juan de Rogel insisted they must (Lyon 1976:202). For one thing, the sacred and the secular formed a seamless garment in Calusa society (Lyon 1976:205). For another, a chief's legitimacy depended upon his mastery of the traditional religious knowledge (Zubillaga 1946:289–290, as translated in Goggin and Sturtevant 1964:192). The Indians who attended Rogel's first catechism lessons "all took off" once the priest ran out of maize to give them, and the situation never really improved during his presence (Zubillaga, as translated in Hann 1991:239, 260–261).

The Spanish presence among the Calusa was to be both short-lived and violent, Carlos's pledges of friendship notwithstanding. Carlos plotted to kill Menéndez on at least three occasions (Solís de Merás 1964:143; Connor 1939:45, 67) and continually harassed the Spanish outpost set up near the Calusa capital late in 1566 under the command of Captain Francisco de Reinoso.

The Calusa attacked a boat party during the establishment of the settlement, killing three Spaniards and bouncing a spear off Menéndez's breastplate (Connor 1939:67). Later, a Calusa party attempted to capture Rogel and make of him a human sacrifice (Zubillaga 1946:607–608, as translated in Lewis 1978:34). Eugene Lyon (1976:201) says "The Spanish were like men besieged in their blockhouse—they dared not leave it without armed guard." The Spanish documents translated by Jeannette Thurber Connor (1939:31–81) tell a tale of virtually uninterrupted Indian hostility, not only by the Calusa but by other Indian groups throughout South Florida.

By the spring of 1567 Reinoso had had enough. He maneuvered Rogel—who opposed violent countermeasures—into going to Havana, then seized and killed Carlos and his closest advisers. In his place the Spanish installed Don Felipe (Zubillaga 1946:296, as translated in Lewis 1978:29). Things seemed to have gotten no better for the Spanish even though Felipe had been to Havana (Connor 1939:39) and took the lead

14

among the Indians in accepting Christianity (Solís de Merás 1964:151). Reilly says his "great reverence" for the cross was merely an attempt to counter Carlos' influence with the Spanish (Reilly 1981:409), and it did not take Rogel long to realize the superficiality of Felipe's conversion (Zubillaga 1946:203, as translated in Hann 1991:244). "He has made me very suspicious that he is not proceeding . . . with as much sincerity as I would like," the priest wrote. Felipe said a lot of the right things, but his actions belied his words. He would not allow his daughter to be baptized when she was seriously ill and kept stalling Rogel on abandoning aboriginal religious practices (Zubillaga, as translated in Hann 1991:244–248, 262).

Felipe was if anything more dangerous because he was a more powerful leader than Carlos (Lewis 1978:31). A good example of his ruthlessness is the way he had fifteen chiefs slain to avert a rebellion, then danced with his supporters around the heads of four of them (Vargas Ugarte 1935:91). Even so, he could not keep four other chiefs from going over to the Tocobaga (Zubillaga, as translated in Hann 1991:262).

When the Calusa in 1569 attacked a landing party sent out by Pedro Menéndez Marquéz, nephew of the conquistador, Menéndez Marquéz retaliated by killing Felipe and twenty of his supporters (Connor 1939:39). Calusa leadership then passed to Don Pedro, a first cousin of Carlos (Fontaneda 1973:31; Zubillaga 1946:309–311, as translated in Goggin and Sturtevant 1964:194). That did not help, either. Pedro also had been to Havana, but would not accept Christianity and "became worse" than he had been before the Spanish first sought his friendship (Fontaneda 1973:31). In the face of yet more hostility, Reinoso ordered the outpost abandoned (Vargas Ugarte 1935:107). Thus ended the formal Spanish presence among the Calusa, only three years after it had begun.

What was this Indian society that proved so impervious to Spanish attempts to control it? It certainly met the anthropological definition of a chiefdom—a society in which a certain kinship group had acquired a preferential access to resources and power—and seemed well on the way to breaking the bonds between heredity and power and thus becoming a state (Kottak

1974:193–194). This last is based upon the irregularity of the succession to power, as compiled by John M. Goggin and William C. Sturtevant.[3]

It also could have been a venerable society. The archaeological evidence suggests that the people inhabiting the Charlotte Harbor-Estero Bay area were extending their influence eastward by A.D. 1400 (Griffin 1988:142) and both northward and southward at least by A.D. 1200 (Griffin m.s.:13; Goggin 1964a:123; Willey 1982:555) and possibly as early as A.D. 800 (Widmer 1988:279).[4] This influence in turn suggests the sort of power possible only in a society where the leaders can compel obedience.

Further, these people appear from the archaeological record to be indistinguishable from the historic Calusa (Milanich and Fairbanks 1980:234). The only reason there is any uncertainty at all is the difficulty of identifying any excavated archaeological site unequivocally with the historic Calusa (Milanich and Fairbanks 1980:243). Only within the past few years has anyone felt confident enough to identify a site as Calusa; Widmer (1988:86–87) does so with regard to a burial mound on Pine Island in Charlotte Harbor. There is a strong presumption that

[3]Goggin and Sturtevant (1964:193–194) chart a Calusa succession by which leadership passed "(1) to the dead chief's brother (Carlos I or Senquene), said to be a usurpation, then (2) to Senquene's son Carlos II, then (3) to Felipe, who was both sister's son and adopted son of the first chief, and finally (4) to Pedro, a 'first cousin' of Carlos II."

[4]Regarding possible expansion of the Charlotte Harbor people by A.D. 1200, both Goggin (1964a:124) and Gordon R. Willey (1982:555) note the presence of Glades pottery types north of Charlotte Harbor, while John W. Griffin (m.s.:14) and John C. and Linda M. Van Beck (1965:18) cite the dominance of plain wares typical of Charlotte Harbor in the Naples-Marco area, a development that to Griffin suggests southward expansion. Alternatively, Marquardt has said, "One could argue that the Calusa complexity . . . was a sudden and recent development, stimulated by European presence." He says leaders may have increased their power and control to counteract the decentralizing influence of prized European goods on the edges of their territory (Hann 1991:xvi).

when Indian and Spanish artifacts are found together in Southwest Florida that the Indian articles were used by the Calusa, but the evidence remains in most cases less than conclusive.

The power wielded by the cacique was evident to Spanish observers. Two Spaniards delivered by east coast Indians to René Laudonnière, founder of the short-lived French presence in Florida, told him that Carlos was "the strongest and most powerful Indian in the country, a great warrior having many subjects under his sovereignty" (Laudonnière 1975:110). The survivors of one Spanish shipwreck were brought to him even though the wreck took place at least 200 miles from his capital (Solís de Merás 1964:221). His house was large enough to hold 2,000 persons, and he received visitors while positioned alone on a raised seat "with a great show of authority" (Solís de Merás 1964:145–146). He wore special ornaments—a gold object on his forehead and bead bands on his legs (Zubillaga 1946:310, as translated in Goggin and Sturtevant 1964:191)—and was greeted in a special manner. The subject would kneel, raising his hands palms up, and the chief would place his hands atop them (Goggin and Sturtevant 1964:192).

The ability to amass and maintain such power had both material and spiritual aspects; Jerald T. Milanich and Charles H. Fairbanks (1980:243) cite "the subsistence potential of both the Southwest Florida coastal waters and the savannahs and wetlands of the Okeechobee Basin and . . . the need to maintain exchange routes," a matter that will be discussed in more detail below. Equally vital, especially in convincing his people that he was the one to follow, was the cacique's mastery of the Calusa religious system, as noted previously. Laudonnière says the two Spaniards told him, "The [Calusa] king was held in great reverence by his subjects and . . . he made them believe that his sorceries and spells were the reason why the earth brought forth her fruit." Once or twice a year he would withdraw to perform secret rites—to observe them meant quick death—and at harvest time a Spaniard would be sacrificed (Laudonnière 1975:110).

What we know of Calusa theology is due largely to the letters of Rogel. The Indians "say that each man has three souls. One is the little pupil (*niñeta*) of the eye; another is the shadow that each one casts; and the last is the image of oneself each one sees in a mirror or in a calm pool of water," he wrote. Rogel said the Calusa believed that only this first soul remained with the body at death and the Indians went to the graveyard to consult with it—or, in Rogel's view, with the devil—in order to know what would happen in other times and places. The deceased "also tell them that they should kill Christians and other evil things. And when someone becomes ill, they say that one of the souls has left him and the shamans go there to search for it in the [woods] and they say that they bring it back, making the same movements as people make who are driving some sheep or goat into a corral that does not want to be shut up. And then they put a great deal of fire at the door of the house and at the windows so that it will not go out again. And they say that they put it back in the man again through the nape of the neck. . . . They have another error also, that when a man dies, his soul enters into some animal or fish. And when they kill such an animal, it enters into another lesser one so that little by little it reaches the point of being reduced into nothing" (Zubillaga 1946:278–281, as translated in Hann 1991:237–238).

"The oneness (*unidad*) of God and his being the creator of every good, they admit to. They also believe those who govern the world to be three persons, but in such a manner that they say the first one, who is greater than the other two, is the one to whom the universal government of the most universal and common things belongs, such as the heavenly movements and the seasons, (*tiempos*), etc. And the second one is greater than the third, that to him belongs the government of the kingdoms, empires, and republics. The third one . . . is the least of all and the one who helps in the wars" (Zubillaga, as translated in Hann 1991:238–239).

Birds and eyes appear to have been important. Eight crested birds with accentuated eyes, fashioned of metal that had to be obtained from the Spanish, are reported from six different sites. One came from Punta Rassa, at the mouth of the Caloosa-hatchee, while the others almost certainly came from either higher up the Caloosahatchee or in the Okeechobee Basin west of Lake Okeechobee, though one is listed simply for "Manatee

County" (which in the nineteenth century reached the west shore of Lake Okeechobee) and another for St. Marks in northwest Florida (Goggin m.s.:579–581). Alice Gates Schwehm (1983:112) points to an emphasis on the eye in prehistoric wood objects from the Okeechobee Basin and speculates that this concept was passed on to the Calusa. This assumption is consistent with Rogel's report that the soul which remains with the body after death resides in the eye.

Whether the Calusa religion was affiliated with the Southern Cult,[5] a late prehistoric manifestation defined by a series of motifs in which the eye is relatively prominent, is problematical. Antonio J. Waring and Preston Holder (1968:9) included the Key Marco site, a late prehistoric site probably occupied by the Calusa, in their list, and Clarence B. Moore (1902:460–461) reported a pendant from another site on the same island that contained eye motifs. But Marion Spjut Gilliland, in her study of the Key Marco material culture, says that "While there are some similarities in form, they seem to be insufficient to assign Key Marco to the Southern Cult" (Gilliland 1975:39). Widmer agrees.[6] (There is controversy as to the age of the Key Marco site and

---

[5]The Southern Cult is defined by Waring and Holder (1968:9) as "a cult complex in association with platform mounds . . . found virtually intact over a wide geographic area, and . . . chronologically late" in prehistory. The criteria they use are eight motifs, three of which (the forked eye, the open eye, the hand and eye) involve the eye (Waring and Holder 1968:10). Their analysis includes two objects from Key Marco, a "baton" that "was about two feet long, was made of wood, and terminated in a 'grooved knob or boss to which tassel cords had been attached'" and a bird painted on wood (Waring and Holder 1968:20, Figs. 3q, 5g). A shell gorget illustrated by Waring and Holder is strikingly similar to two reported from Key Marco (Waring and Holder 1968:Fig. 4b; Gilliland 1975:Plates 113–114).

[6]"What we see then in South Florida is a distinct ceremonial complex, perhaps best termed the South Florida Ceremonial Complex, maintaining a distinct regional and artifactual expression ultimately derived from a common pan-eastern Hopewellian-based ceremonial complex, but incorporating various Southeastern Ceremonial Complex motifs and traits into an indigenous religious system" (Widmer 1989:179–180).

19

whether the objects found there can be tied to the historic Calusa, points that will be discussed in detail in Chapter 5.)

As noted before, human sacrifice was an important part of Calusa rituals. "Each time a son of the cacique dies each inhabitant sacrifices his sons or daughters. . . . When a chief himself or the chieftanness dies, they kill his or her own servants. . . . Each year they kill a Christian captive so that they may feed their idol" (López de Velasco, as translated in Hann 1991:316). The sacrifices presumably took place in temples atop some of the mounds that dotted the Calusa area, mounds that today are recognized to have been built not by the historic-period Calusa but by predecessor cultures (Milanich and Fairbanks 1980:27).[7]

Rogel speaks of "a temple of idols . . . which were some very ugly masks" (Zubillaga, as translated in Hann 1991:287). The best idea of their appearance probably comes from the score or so wooden masks reported for Key Marco (Gilliland 1975:Plates 39–58). These do not look all that ugly, but then we are not seeing them worn by Indians who had "conceived a great hate" of the observer "because he had revealed their secrets and profaned their religion" (Zubillaga, as translated in Hann 1991: 287).

There appear to have been processions, idols, and charnel houses. The Tocobaga defleshed the corpses of leading men and reassembled the bones, while the Tequesta would remove the large bones from a leader's body and keep them for worship

---

[7]The assumption that the Calusa performed sacrifices in temples atop mounds is based on the observation of Rogel at the Calusa capital. He reports that the temple or temples atop the Calusa mound were used to store masks, perhaps of the type found at Key Marco (Zubillaga 1946:607–608, as translated in Goggin and Sturtevant 1964:199; Gilliland 1975:Plates 39–58). Ortiz reported a charnel house for the Tocobaga (Elvas 1984:150; Garcilaso 1980:66). As for the mounds themselves, they once were believed to have been the work of the Calusa (Goggin and Sturtevant 1964:194–197), but that view has been abandoned, though it is recognized that the Calusa undoubtedly added to them (Milanich and Fairbanks 1980:246) and definitely used them, as shown by the number of the distinctive tablets that were found in association with mound burials (McGoun 1981:23).

20

while burying the rest of the body (López de Velasco, as translated in Hann 1991:318–319). The Calusa quite possibly had similar practices.

The Spanish felt Calusa power from the first. Officially, the first European visitor was Juan Ponce de León in 1513, but it is virtually certain there had been previous visits. The most dramatic evidence in that direction is that Florida is indicated on a map dating from 1502 (Smith and Gottlob 1978:1).

Ponce met an Indian who spoke Spanish (Davis 1935:20) and found gold at Charlotte Harbor (Swanton 1979:101; Davis 1935:44). The case here is not very strong, because both the gold and the Indian could have come from the Antilles.[8] The canoes in which the Calusa came forth to meet Ponce, "some fastened together by twos" (Davis 1935:20), could have made the trip to the islands, and J. Leitch Wright, Jr., believes such trips were made before the time of Menéndez (Wright 1981:45). Fontaneda said that "many Indians from Cuba" arrived in Calusa territory in the time of the first Carlos and that he "made a settlement of them, the descendants of whom remain to this day [1575]" (Fontaneda 1973:29).

Further indication of previous contact was the hostility. The Indians who came to meet Ponce's party in canoes did not constitute an aboriginal Welcome Wagon. "Some went to the anchors, others to the ships, and began to fight from their canoes. Not being able to raise the anchors they tried to cut the cables" (Davis 1935:20). After nine days of off-and-on hostilities, in which one Spaniard and several Indians were killed, Ponce withdrew (Barcia 1951:2; Davis 1935:20). T. Frederick Davis (1935:41) places these events at Charlotte Harbor in the water near Pine Island.

---

[8]Metals used by the South Florida aborigines would have been carried a long distance if they were of North American origin, but it is far more likely they were obtained from the Spanish. There were some silver and copper in the Lake Superior area, and "a very small amount of gold was used by the mound-building tribes" (Swanton 1979:494–495), but it is virtually certain that all Florida gold and silver came from Central and South America via Spanish ships.

Ponce would fare far worse when he attempted to establish a settlement among the Calusa in 1521, again presumably in Charlotte Harbor. Davis interprets the Spanish accounts as saying that at least eighty of his party either were killed outright or died of their wounds. Ponce himself was in the latter category, succumbing in Cuba to the effects of an arrow wound in the thigh (Davis 1935:61–63).

The 1517 Cordova expedition to Mexico likewise had an unpleasant time when it revisited the Ponce landing place in search of water. The Indians attacked as the Spaniards prepared to return to their ship. A score of Indians were killed, and at least half a dozen Spaniards wounded. One Spaniard was carried off alive, quite possibly winding up as a sacrifice (Díaz de Castillo 1956:15–16).

All of this evidence is a strong argument against the view of such writers as Warren H. Wilkinson (1947) and Rolfe E. Schell (1966) to the effect that Hernando de Soto made his 1539 landfall at Charlotte Harbor rather than Tampa Bay, the site more commonly accepted. Why would de Soto, who had to have known about the Ponce and Cordova visits, choose the site of such hostility for landing? While there was one skirmish on the beach, in which two Indians were killed, and two Indian raids in which three Spaniards were taken captive (Lewis 1984:146, 148; Garcilaso 1980:60, 90, 92), de Soto did not face nearly the resistance that had been unleashed against Ponce and would be directed against Menéndez.

Of course, de Soto had the advantage of numbers. While Menéndez first went ashore with 30 men, the initial de Soto landing involved 300 (Garcilaso 1980:60). A stronger argument concerns not the Indians de Soto met but the European he found: Ortiz. The fact that Ortiz, who had gone to the landing site of Pánfilo de Narváez's 1528 expedition to look for Narváez, was rescued by de Soto (Lewis 1984:149; Garcilaso 1980:62–63, 75) indicates that both expeditions landed in the same area. Had those landings been among the Calusa, then the maize fields near the landing place (Hodge 1984:21) would have had to be at the north end of Charlotte Harbor. If this were the case, it is inconceivable that Fontaneda, who had been among the Indians for seventeen years, would not have known about it. Fontaneda (1973) does not even mention maize.

Henry F. Dobyns (1983:126–131) accepts Charlotte Harbor landfalls, arguing for a protocontact Calusa population of nearly 100,000. This extraordinary figure far outstrips the previously highest estimate of 10,000 to 15,000 for the Southwest Florida coast and 5,000 to 10,000 for the Lake Okeechobee Basin at European contact, given by Milanich and Fairbanks (1980:246). Two critical assumptions underlie Dobyns's conclusion. The first, which seems valid, is his contention that previous research had underestimated severely the effects of disease upon Indian groups even before they came into direct contact with Europeans. The only other person really to come to grips with this question is Wright, who says (1981:51) the Calusa already had lost population before the time of Menéndez, though he does not cite any figures as high as the 80 percent used by Dobyns (1983:287).

The second, far more questionable, is that of maize horticulture. How could it have been overlooked by the authors of all the primary documents on the Calusa (Goggin and Sturtevant 1964:184)? As late as 1675, Bishop Gabriel Díaz Vara Calderón said South Florida Indians were "living only on fish and the roots of trees" (Wenhold 1936:11–12). In 1697 the Calusa, upon seeing hoes brought by a Spanish party, "asked why they had not brought blacks who might dig with the hoes, because the Indians did not know how to" (Romero, as translated in Hann 1991:184–185). In a recent reappraisal of cultivation and the Calusa, Milanich (1987:180) concludes that "the jury is still out." He doubts Dobyns's conclusions but says they cannot be ruled out in the absence of further research. William H. Marquardt agrees, saying (1986:66), "The question of Calusa horticulture is still open."

The Milanich-Fairbanks population estimate seems closest to the mark. Given the numbers of Indians reported by Solís de Merás, the 4,000 to 7,000 estimate of Goggin and Sturtevant (1964:187) seems too low, and the Dobyns figure is insupportable in the absence of cultivation in an area as small as the Calusa heartland.

As for settling the landfall question through the physical descriptions as given in accounts of the de Soto expedition, "it is only possible to identify a general region" (Brain 1985:xiii). The best case for Charlotte Harbor comes from the account of Garcilaso de la Vega, who says (1980:60, 95, 116, 121) the landing place was forty-seven leagues from a village called Ocali, which undoubtedly was in today's Marion County (Milanich 1978b: 69). Depending on which of the many different "leagues" is meant (Haggard 1941:78–79), that translates to somewhere between 122 and 171 miles, which correlates more to Charlotte Harbor than to Tampa Bay. This estimate of course assumes that the Spaniards were traveling consistently in the same direction and were able to know precisely how far they had traveled, two very dubious assumptions. Further, the account of the Gentleman of Elvas appears to suggest a shorter distance between the landfall and the village he calls Cale (Lewis 1984:146–155). His estimation is significant because he was a member of the expedition and Garcilaso was not (Fernandez 1975:85).

Other contemporary Spanish accounts also support a Tampa Bay landfall. A navigator's guide prepared before the time of de Soto gives a description of Bahia Honda that matches Tampa Bay, and Luys Hernandez de Biedma, another eyewitness chronicler, identifies that bay as the landfall. Yet another expedition member, Rodrigo Ranjel, says the landfall was due north of the island of Tortuga, which fits Tampa Bay far more than Charlotte Harbor, and differentiates it from the Bay of San Juan Ponce, generally interpreted as being Charlotte Harbor (Chaves 1983: 366; Biedma 1904:3; Oviedo 1904:54). Recent research on objects found in the Tampa Bay area "leave no doubt that DeSoto landed at Tampa Bay and that his army left behind ample archaeological evidence of its presence. Comparable evidence is not found anywhere else in Florida" (Milanich 1989:300).

Further, inasmuch as the Calusa had to be known to de Soto due to Ponce's voyages, why would they not have been mentioned in any of the chronicles had the expedition in fact landed among them? This is one of the major points made by John R. Swanton (1985:138) in support of Tampa Bay. There is still enough doubt to hearten the champions of Charlotte Harbor. Jeffrey P. Brain leaves the question open after his reanalysis of the data (1985: xiii). Louis D. Tesar argues that "no conclusive evidence has yet been presented" in support of Tampa Bay (1989:279), and Lindsey Williams argues for a landfall on the north shore of Charlotte Harbor (1989:280–294), but the weight of the evidence continues to favor Tampa Bay.

The best single source for first-hand information about the Calusa of the Menéndez period is Fontaneda's memoir. Fontaneda paints an expansive picture of Calusa influence. He says Carlos held sway over at least fifty towns ranging from Tampa Bay to the Keys and inland to Lake Okeechobee. He does not claim any direct Calusa control over east coast groups, but does tell of Carlos redistributing booty brought to him by the Ais, reserving "what pleased him, or the best part" for himself (Fontaneda 1973:30–31, 34). The claim of Calusa control over the Lake Okeechobee area is reinforced by Rogel's report that disgruntled Calusa frequently threatened "to leave us and go to the lakes to live" (Zubillaga, as translated in Hann 1991:240).

As noted previously, redistribution between the coast and the Lake Okeechobee Basin was especially important. "For instance, should fishing be bad one week at Calos (the name frequently given for the Calusa capital), dried palm berries and smilax or zamia flour (or unprocessed roots) could be already on hand or brought by canoe from a village on Lake Okeechobee via the Caloosahatchee River." Alternatively, "smoked fish brought by canoe from a village elsewhere on the coast could feed the Calos villagers" (Milanich and Fairbanks 1980:243). Sturtevant would dispute the assertion about smilax or zamia. "There is no evidence for the use of zamia (or smilax) by the pre-Seminole Florida Indians; there are reasons for supposing the Seminole use of zamia may have been learned from whites from the West Indies" (Sturtevant 1960:12).

Not everyone accepts Fontaneda's belief that the coastal people controlled the basin. Milanich and Fairbanks (1980:189) note that historic artifacts are more common in the basin than on the coast and conclude that "If such European items were,

25

indeed, valued by the aboriginal populations, this would indicate that the inland peoples controlled the networks along which such goods were traded." William H. Sears (1982:200–201) says that while Fort Center, near Lake Okeechobee on Fisheating Creek, "was part of the . . . Calusa empire . . . I see no reason to believe that the Calusa . . . ever came inland to Fort Center."

I have argued elsewhere (McGoun 1981:36–37), with regard to the distribution of certain distinctive tablets, that the data are skewed by extensive pot-hunting on the coast. The same argument applies to other European objects, inasmuch as most were fashioned from precious metals. Taken in conjunction with the archaeological evidence for greater population on the coast, and the Spanish accounts of Calusa power, the case for coastal supremacy seems to be persuasive.

Some European sources are even more generous than Fontaneda in their estimate of Calusa power. Juan Fernández de Olivera is quoted as saying Calusa influence reached up the Atlantic Coast almost to St. Augustine in 1612 (Goggin and Sturtevant 1964:187), and both Frederick Webb Hodge (1911: 195) and W. E. Safford (1919:426) believe that the Calusa held sway as far as Cape Canaveral. Laudonnière (1975:71–74) speaks of an alliance between the Calusa and the Cape Canaveral Indians.

The Olivera estimate may be too generous, but the concept of at least some Calusa influence in the Cape area is reinforced archaeologically by discovery of two crude examples of the South Florida tablets at the Gleason Mound, in the Cocoa area (see Figure 3). I feel certain these objects were personal symbols of political power and that therefore their presence indicates Calusa political influence, perhaps minor and perhaps transient but influence nevertheless (Rouse 1951:199, Plate 7; McGoun 1981:35).

As Milanich and Fairbanks note (1980:26), there is a transitional area on the east coast as well as on the west. The last pottery type to emerge in prehistoric South Florida, a type called Glades Tooled because of the rim treatment, is not found in the Cape Canaveral area (Rouse 1951:254), and Jonathan Dickin-

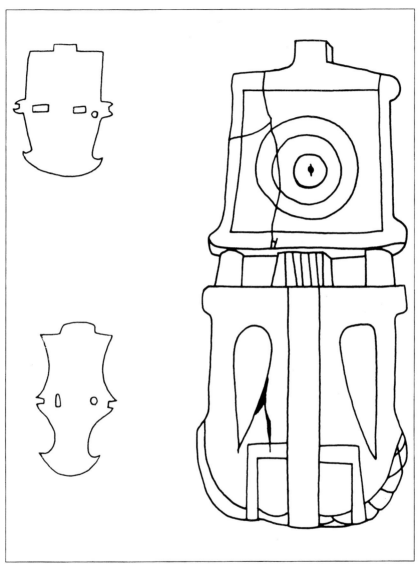

Figure 3. Tablets from (*left*) Gleason Mound (Rouse 1951:Plate 7) and (*right*) Key Marco (Gilliland 1975:Plate 35)

son, in recounting his passage through the area in 1696, does not mention either encountering or hearing about the Calusa (Dickinson 1981).

The historic inhabitants of the area were the Ais and Jeaga, first identified by Fontaneda (1973:32). Fairbanks (1957:33, 39) places the Ais territory as running southward from a point just north of Cape Canaveral to St. Lucie Inlet, with the Jeaga to the south. Maps with the Dickinson journal indicate a settlement called Hoe-Bay, presumably Jeaga, on the south side of Jupiter Inlet and another called Jece, apparently Ais, on the barrier island just south of Vero Beach (Dickinson 1981:Map 1). Both groups lived along the coast, though they ranged inland (Milanich and Fairbanks 1980:238). Fairbanks (1957:35) believes the Ais territory extended inland to Lake Okeechobee at its southern end. These were smaller societies than the Calusa. Milanich and Fairbanks (1980:239) estimate their combined population at 2,000. They also were poor, according to Fontaneda (1973:32).

Like the Calusa, they demonstrated from the first a dislike for the Spanish. When Ponce de León in 1513 attempted a landing that Davis locates north of Palm Beach (Charles D. Higgs [1942:37] thinks it was at the Ais capital), the Indians "immediately tried to take the boat, the oars, and the arms. . . . they, with their arrows and armed shafts—the points of sharpened bones and fish spines—wounded two Spaniards" (Davis 1935:18, 48). Menéndez Marquéz tells of two massacres on the Atlantic coast in the 1560s (Connor 1925:41).

Menéndez de Avilés got a better reception, possibly because he gave the Ais cacique "many presents and gave him many things for barter, and did likewise to his principal Indian men and women" during his visit in November of 1565. When Menéndez departed for Cuba he left behind a garrison numbering more than 200 Spanish soldiers and French captives, under the command of Captain Juan Velez de Medrano. Menéndez took the precaution of locating the garrison some distance from the main Ais village to lessen the possibility of conflict (Solís de Merás 1964:128).

This would be a very short-lived effort at settlement. Four

months later, when Menéndez was returning to St. Augustine, he came across the supply ship sent to deliver food to the Florida forts and found it manned by mutineers from the Ais garrison. All but seventy-five of the party apparently had starved to death (Lyon 1976:150). The troops were relocated among the Jeaga, probably near Jupiter Inlet, but within a year the effort had been abandoned (Lyon 1976:139–141). Rogel tells of an Indian attack upon this garrison (Vargas Ugarte 1935:83).

The scant mention given the Jeaga in sixteenth-century reports indicates that the Ais were the stronger of the groups then, and that appears to have been the case in Dickinson's time as well. Charles Andrews reaches this conclusion (1943:40) on the basis that they were able to appropriate part of the loot from the Dickinson party's shipwreck, even though the wreck occurred in Jeaga territory (Dickinson 1981:34) and that Dickinson describes better buildings for them than for the Jeaga.

According to Dickinson, Hoe-Bay consisted of "little wigwams made of small poles stuck in the ground, which they bended one to another, making an arch, and covered them with thatch of small palmetto-leaves" (Dickinson 1981:11), whereas the house of one Ais cacique "was about forty foot long and twenty-five foot wide, covered with palmetto leaves both top and sides" and that of another is cited as being large (Dickinson 1981:23, 29). Additionally, Dickinson (1981:39) tells of an Ais cacique, probably the paramount leader, receiving "great baskets of dried berries . . . from divers towns."

Like the Calusa, neither group had horticulture. Dickinson (1981:47) reported no crops on his northward journey until he reached the Cape Canaveral area, and those were pumpkins, possibly grown by intruders from the north.[9] Dickinson says

[9]The evidence that the pumpkins Dickinson saw at Cape Canaveral could have been grown by intruders from the north is both historical and archaeological. Attacks into Florida from English territory to the north had begun in earnest by 1685 (Matter 1975:31), a decade before Dickinson was shipwrecked, while pottery of a North Florida type, San Marcos, has been found at Sebastian Inlet in a context suggesting it was brought south by migrants rather than traded (Rouse 1951:217).

29

(1981:13) that the Jeaga fished with spears. He alludes (1981:21) to fishing by the Ais and says the latter group gathered cocoplums, sea grapes, and palm berries (1981:30). At the principal Ais village he saw a chunk of ambergris, a waxy substance believed to originate in sperm whales, that he estimated at five pounds in weight (Dickinson 1981:43). Evidently, neither group had a secure food supply. Dickinson (1981:36) reports a scarcity at the principal Ais village in cold weather, and the authors of a Jupiter Light station survey believe the village at Jupiter Inlet was occupied only seasonally (Weed et al. 1982:25).

Dickinson (1981:19) tells of an Ais sailing canoe, the sail probably reflecting Spanish influence. The use of black drink (Dickinson 1981:25), a custom found throughout the aboriginal Southeastern United States, suggests that the trait also was present in South Florida. One notable difference between Menéndez's time and Dickinson's, 130 years later, is that the Indians had changed in their attitude toward the Spanish. Dickinson's journal indicates repeatedly their preference for the Spanish over the English, perhaps reflecting their knowledge that there would be retribution if they mistreated a Spaniard.

To the south of the Jeaga—both Goggin (1940:274) and Fairbanks (1957:39) put the boundary in northern Broward County—were the Tequesta, clearly the second most powerful contact-era Indian society in South Florida after the Calusa. Their power is indicated by the fact that on one occasion they received as a present from the Ais two Spanish captives (Connor 1925:41). Their population probably was upwards of the 2,000 estimated for the Ais and Jeaga, but not too much upwards. The frequently cited total of 1,000 for all three groups (Fairbanks 1957:41) seems unrealistically low.

Calusa relations with the Tequesta seem to have varied over time. The Tequesta appear to have been close to the Calusa politically and ethnically—Solís de Merás (1964:210) says the Tequesta cacique, whom the Spanish also called Tequesta, was related to Carlos—but distant enough geographically to assert a degree of independence. On one occasion, the Tequesta chief, who previously had been subservient to Carlos, refused a demand that he turn Spanish captives over to the Calusa and fought off an attempt by the Calusa to take them by force (Solís de Merás 1964:222). There probably was a considerable amount of no-man's-land between Tequesta and Calusa territory south of Lake Okeechobee, given the relative paucity of both resources and population in the intervening Everglades. In fact, much travel between the two areas probably was made by sea.

The Tequesta also were subject to a Spanish mission presence during the Menéndez period, and at first it seemed as if the experience here would be far different from that among the Calusa. In January of 1568 Brother Francisco de Villareal, the cleric in charge of the Tequesta mission, wrote to Rogel that "I am instructing the Indian children up to fifteen years old. The others do not come, not because they refuse to become Christians, but because they find it so difficult to learn. . . . The young chief is favorably disposed toward the Christians and I think he will remain so" (Vargas Ugarte 1935:76).

Amid this optimism, however, Villareal notes that hunger was a problem. "For a few days we were without food. . . . Then most of the Indians went off to an island about a league away to eat nuts and dates. . . . For the last two or three months they were all so hungry that they stopped coming for instruction and came instead for food" (Vargas Ugarte 1935:76). The Jesuit could be of little help here. "I went to Havana. . . . I brought back some food, though not much as I had no ship in which to bring it" (Vargas Ugarte 1935:76). Perhaps as was the case with the Ais garrison, tensions were generated by the strains of feeding nonproducing visitors in a nonhorticultural society.

On the other hand, Villareal may have overestimated the Indians' acceptance of his faith. He reports that shamans were unhappy about his intercession with a sick girl who later died and wondered, in a phrase reminiscent of Rogel's observations about the Calusa, if the willingness of the sick to be baptized grew "out of fear or lack of understanding or for the maize they like so much" (Vargas Ugarte 1935:77–78).

The maize evidently had been brought by the Spanish from Havana. The Tequesta lived off the land and the sea; besides their gathering of nuts and dates, Villareal mentions "whale

meat and fish" (Vargas Ugarte 1935:76). Analysis of plant remains from the Granada site, believed to be close to if not part of
the mission site, indicates occupation only in the fall, "during
which time a narrow spectrum of plant resources was exploited,"
consisting mostly of fruit. The Granada researchers believe,
however, that fruits were merely a supplement to a diet based
mostly on animals (Griffin et al. 1982:3, 245–246). A site on New
River in downtown Fort Lauderdale has been identified as a
shark-butchering area (Graves 1989:255–256).

Matters came to a head within weeks of Villareal's optimistic
report, when the Spanish killed an Indian leader. The Tequesta
surrounded the outpost, killing four of its inhabitants, and
probably would have killed the rest had not Menéndez Marquéz
arrived "at the very hour they were in the fight" and evacuated
the survivors (Lyon 1976:203; Connor 1925:39). Interestingly,
three months later Rogel insisted, in a letter to the future St.
Francis Borgia, head of the Jesuits, that "the natives are peaceful,
tractable and well disposed toward the Faith. . . . They are very
different from the natives of Carlos who are energetic, turbulent
and intractable" (Vargas Ugarte 1935:81). Evidently he had not
yet heard of the uprising.

The Keys Indians are perhaps the least known of the South
Florida groups at European contact. Both Solís de Merás
(1964:188) and Fontaneda (1973:31) say the Calusa exercised
control there. In fact, Carlos may have gone personally to the
site of a Spanish shipwreck on the Keys (Laudonnière 1975:110).
However, other Spanish reports place the area under the control
of the Tequesta (Connor 1925:59) or of "a cacique they call
Matecumbe" (Connor 1925:51). Goggin (m.s.:48, 74) says that
from an archaeological standpoint the Keys sites "show a great
similarity both in construction and contents" to the Tequesta
area and concludes that they were part of that area "culturally, if
not always politically."

Given the isolation of the individual islands, the lines of
control probably varied considerably from place to place and
time to time. Like the other South Florida coastal groups, the
Keys Indians depended heavily upon sea resources. Fontaneda
(1973:26) says fish—notably tuna—plus turtle, mollusks, and

whales were eaten by all the Indians and seals by the leaders. They may have eaten well but, as Purdy (personal communication) points out, they hardly had tuna or seals. Again like the other South Florida groups, they were hostile to the Spanish, at least in the early years. Menéndez de Avilés tells of the killing of eight Spaniards on a southbound boat (Connor 1925:33).

The Keys Indians, along with the east coast groups, almost certainly were not as highly organized as the Calusa. Nevertheless, the presence of whale hunting indicates there was at least some organization. As Stephen L. Cumbaa puts it (1980:9), "There are important social ramifications in the cooperative group effort necessary for the spotting, capture, butchering, distribution and redistribution of the tremendous amounts of meat and blubber present in even one large individual."

In all probability, however, none of the Keys leaders had any authority beyond his own village. This limitation in turn relates to the environmental factors that set the Calusa apart, specifically the rich marine resources of the Southwest Florida mangrove coast. The Calusa's cultural complexity is especially notable in that it seems to have been established and maintained in the absence of horticulture. Fontaneda (1973:30) speaks of a "bread of roots" eaten in the Lake Okeechobee area, and Laudonnière (1975:111) mentions a "bread" at Serrope, a town of uncertain affiliations that probably was on Lake Kissimmee, though most sources identify it as being on Lake Okeechobee,[10] but it is unclear if the roots were cultivated. Aside from that, there is no evidence for horticulture in the area under Calusa control in the time of Menéndez.

That, however, did not matter. The Calusa coast was host to

[10]The most comprehensive departure from the view that Serrope was Lake Okeechobee is offered by Stephen Hale (m.s.). His case, a strong one, is based on the fact that the account given in Laudonnière regarding capture of a bride meant for Carlos fits Lake Kissimmee much better than Lake Okeechobee and also that the Indians around Lake Okeechobee were known to be subject to the Calusa and would not have dared carry out the abduction.

such large numbers of species that some were available at any given time (Widmer 1988:237). And then there were the shell-fish; Goggin (m.s.:26) says "The clam beds of the southwest coast are said to be the largest in the world" and "at the mouths of many streams oyster reefs thrived." Scallops and various species of conch also were important (Goggin and Sturtevant 1964:186). Shellfish are not a major contributor to the diet of coastal people, Widmer says (1988:248–249), but they are important in that they are available year-round and can be collected by all members of a society, including children, with relatively little effort.

"Probably in no other non-agricultural North American communities, except those of the North Pacific Coast, was a greater food supply available with such ease," Goggin asserts (1964a: 122). Solís de Merás (1964:148) speaks of "many kinds of very good fish," plus oysters, at a banquet Carlos threw for the Menéndez party. Fish were plentiful elsewhere in South Florida, but evidently there was not a sufficient range of foods to support year-round settlement. Both archaeological and historical evidence indicates that even the Tequesta had to move around in search of food (Griffin 1982:245–246; Vargas Ugarte 1935:76).

Carlos cemented his political alliances with marriages, frequently to a woman from a town subordinate to him (Zubillaga 1946:310, as translated in Goggin and Sturtevant 1964:189). In this sense it appears that his insistence that Menéndez take his sister as a bride was symbolic of his subordination to the Spanish, though the sincerity of the gesture seems to be questionable given Carlos's conduct in general. There probably is no single answer as to why the other villages should have been willing to submit. The Calusa undoubtedly had more to trade than did the poorer inland groups. On the other hand, the various attacks on the Spanish indicate that Carlos was able to mount military expeditions of a power other groups would challenge at their peril. Felipe demonstrated his power when he killed the fifteen rival chiefs (Vargas Ugarte 1935:91). Also, the wives presented to the Calusa leader by subordinate villages (Zubillaga, as translated in Goggin and Sturtevant 1964:189) may have served as hostages.

There were some instances of defiance. Besides the above

noted dispute with the Tequesta, there is the Laudonnière report (1975:112) that the people at Serrope intercepted a bride intended for Carlos and kept her. Such incidents, however, apparently occurred only when the rebels were at a good distance from Carlos.

Widmer (1988:216–217, 279) argues that in understanding the rise of the Calusa it is important to remember that while the Southwest Florida marine resources were rich they also were circumscribed and that a highly organized society was necessary to protect against shortages. He believes the expansion of Charlotte Harbor influence about A.D. 800, as indicated in the archaeological record, is coincident with the Calusa reaching the carrying capacity of their homeland. The Widmer concept presupposes that Calusa society had reached its historic form by A.D. 800. Marquardt (1986:66–67) questions the validity of that assumption in light of known environmental changes during that period. He does not reject Widmer's ideas, but argues that the case is not conclusive.

Whatever may have been the exact date, at some point redistribution became important to the Calusa, in more ways than one. Besides providing access to booty and power, it also opened access to alternative resources as an added safeguard against shortages. Marine resources were used for food only near the coast (Wing 1977:54), but such durable items as shark teeth and shells were available for trade. Widmer believes (1988:275) they could have been exchanged for root crops from the interior, though he suspects trade was more likely with the east coast. On top of that was the aforementioned trade in captives. And once the Spanish—or, more correctly, the Spanish metals—were on the scene, the trade network incorporated another dimension. Goggin (m.s.:560–588) listed an extensive inventory of metal objects and others have added to it. Fairbanks (1968:105) cited the distribution of coin beads as evidence of east-to-west trade.

One especially notable feature is the spread of the distinctive tablets. They have been found, usually in connection with burials, from Charlotte Harbor east to Lake Okeechobee and up to the Kissimmee River and Fisheating Creek systems. Specimens have been found as far north as the Thomas Mound,

35

southern Hillsborough County, on the west coast and the Gleason Mound on the east (Willey 1982:124; Rouse 1951:201, Plate 7).

These tablets, a representation of which appears frequently on either the cover or the title page of the *Florida Anthropologist* journal, have been described variously as representing a crocodilian, a spider, or a jaguar. They "are essentially rectangular, although the lower portion frequently is rounded and somewhat bulbous. They almost always are divided into two zones by a pair of perforations a bit above center. The upper zone frequently is distinguished by concentric circles and/or straight lines in a cross pattern, the lower zone by two 'teardrops' and a geometric motif resembling a stylized mouth." As I noted previously, they surely indicated the extent of Calusa political influence at a specific time, though they also may have had religious significance (McGoun 1981:6, 12, 35).

Such influence need not have been permanent. In all likelihood the zone of Calusa control waxed and waned, reaching its maximum extent in the Thomas and Gleason areas. Both sites also are near the limits of the Glades gritty pottery, and the Gleason mound further is identified with the historic Ais capital (Willey 1982:555; Rouse 1951:254, 201).

Over the years various people have come up with various ideas as to what the objects represent. Douglass (1890:14) saw Christian symbolism reflecting missionary influence. Griffin (1946) sees a spider, Cushing (1973:427) and Goggin (m.s.:596) a crocodilian, and Sears (1977:9) an Olmec jaguar. I have argued recently that the term *chaguala* in the literature may be a Calusa combination of two Spanish words, referring to a caterpillar that attacks the green corn (McGoun 1991). If this speculation is true, it would suggest cultural continuity between the Calusa and people living on the west side of Lake Okeechobee 2,000 years previously.

The Calusa evidently had an extensive hierarchy of offices, though it is hard to know the exact structure in light of the multiplicity of names found in various Spanish sources. Rogel speaks of "The king and his nobles and captains" (Zubillaga, as translated in Goggin and Sturtevant 1964:190), while other sources are quoted as talking of local headmen, principal men, priests, sorcerers, captain general, second captain, and "chief queen," or "principal wife" (Goggin and Sturtevant 1964:190–191).

While some of the citations may indicate merely different names for the same office—for one thing, Rogel notes that "nobles and captains" or "principal men" at the Calusa capital had both religious and political duties (Goggin and Sturtevant 1964:190)—nevertheless they indicate the presence of a well-organized society, one that was heir to millennia of human occupation. Now for the beginning of the story.

TWO

On the Trail of Big Game

The Paleoindian Presence in South Florida

To most people South Florida was still a new land in 1913. There had been Euro-American settlement along the Indian River since the 1840s, but the area had not really become accessible until the Florida East Coast Railway was built in the 1890s. Officials of the Indian River Farms Company decided 1913 was a good time to build and laid out the first plat in what today is the city of Vero Beach. As with most real-estate development in South Florida, one of the first tasks was drainage. As a result of that task, South Florida never again would be so young. In cutting the Main Canal, workers came upon an ancient buried streambed that contained the bones of both human beings and extinct animals (Hrdlicka 1918:23–25). If all the bones were in fact of the same age, that meant that humans had been in the area at least 10,000 years before.

In 1913 this was startling news. The idea that American Indians originated in Asia was just beginning to be accepted, and the chief proponent of that view, Ales Hrdlicka, believed that human beings had reached their present physical form only within the last 15,000 years and could not have crossed the Ice Age land bridge between Alaska and Siberia prior to then (Hrdlicka 1918:37). If such assumptions are accepted, it follows that human beings hardly could have reached Florida 10,000 years ago. Hrdlicka insisted that 10,000-year-old human bones would have to show some difference as compared to contemporary human beings and that the Vero skeletons had in fact been buried at a much later date. He discounted reports that the stratum above the bones was undisturbed by saying, "In old graves, except under unusual conditions, all signs of disturbance of the ground are absent or obscured" (Hrdlicka 1918:36, 49). Given Hrdlicka's preeminence in the field and his position

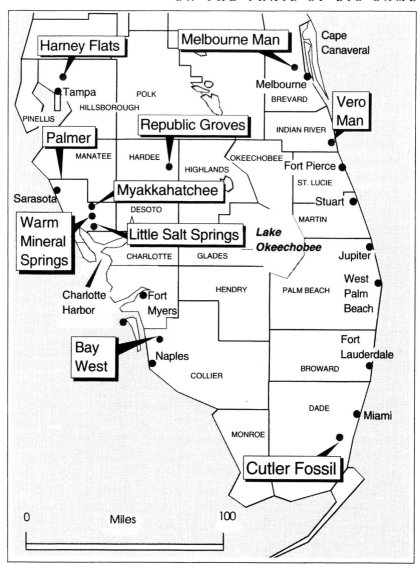

Figure 4. Major Sites of the Paleoindian Period

40

as curator of physical anthropology for the U.S. National Museum, few were prepared to challenge his views.[1]

Some did, however. One of them was E. H. Sellards, the Florida state geologist who first had reported the Vero finds. Another was T. D. Stewart, Hrdlicka's successor at the National Museum, who in 1946 argued that both the single skull recovered at Vero and another skull found at Melbourne in 1925 conformed more to the "long-headed" form known from that early period than to the "broad-headed" form of later inhabitants, including the historic Indians (Sellards 1916:159–160; Stewart 1946:21). Stewart, of course, benefited from the fact that by 1946 a greater antiquity was generally accepted for both the emergence of modern *Homo sapiens* and the peopling of the New World.[2] To say, as Stewart (1946:22) does, that the skulls *could* date from the Ice Age is not, however, the same thing as saying that they *do* date from that period. While W. A. Cockrell and Larry Murphy (1978:2–3) accept such antiquity, Irving Rouse (1951:162, 236) does not. To John W. Griffin (1952:322) and to Jerald T. Milanich and Charles H. Fairbanks (1980:5), the question remains open.

Such is not the case with the finds in two water-filled sinkholes north of the present-day town of North Port Charlotte in the Myakka River valley. There seems no question but that these sites do in fact date from the late Ice Age—or, to use its scientific

[1]In his consideration of the Vero finds, Hrdlicka made clear repeatedly his view that Sellards was not qualified to make conclusions as to their antiquity. Near the front of his report (1918:35) he said, "It is particularly regrettable that in the Vero case anthropologists could not have had the opportunity of examining the evidence . . . *in situ*" (emphasis is his). And he concludes (1918:60) by saying "Our colleagues in collateral branches of science will be sincerely thanked for every genuine help they can give anthropology: but they should not clog our hands."

[2]The gist of Stewart's argument in rebutting the Hrdlicka conclusions is that the Melbourne skull was initially interpreted as being broad-headed because the fragments were placed by Hrdlicka over a mass of clay which "mainly determines the form of the total reconstruction." Stewart instead used temporary wire braces and then crack filler, so that "In this second version the shape of the whole has been determined primarily from the union of the parts" (Stewart 1946:5–6).

name, the terminal Pleistocene Epoch, as opposed to the Holocene Epoch which followed it and in which we live.

A human skull containing what appeared to be brain matter was reported from Warm Mineral Springs in 1960, but it would be another decade before the significance of the find was established. In 1972 two specimens of human bone were found in association with organic material that was dated at approximately 10,000 years of age through the radiocarbon process (Royal and Clark 1960:285–287; Clausen et al. 1975:3). Later in the decade more evidence would come from the nearby Little Salt Spring, in the form of an extinct giant tortoise that evidently had been killed by a 12,000-year-old wooden stake (Clausen et al. 1979:609).

In the 1980s came still more finds. Projectile points of a type thought to be 9,000 to 10,000 years old were discovered at the West Coral Creek site on the Cape Haze Peninsula, roughly ten miles south-southwest of the two springs (Hazeltine 1983:98). Another point of that type and part of one that could be even older were reported for the Myakkahatchee site, several miles north of the springs (Luer et al. 1987:146). A knife and a core that could be late Paleoindian were found at the Republic Groves site, upstream from the springs in the Peace River valley (Wharton et al. 1981:77).

Still, all of these finds were on the fringes of the area generally defined as South Florida. Had Ice Age inhabitants stopped here, rather than proceeding southward? That question would not be answered until a dry sinkhole in Dade County south of Miami was systematically examined beginning in October 1985. One portion of the Cutler Fossil site produced both Ice Age projectile points and a hearth containing charcoal that has been dated by the radiocarbon method at nearly 10,000 years of age. Another produced human bones in association with bones of extinct animals, including horse and condor (Carr 1986:231–232).

All of these finds relate to the culture that archaeologists call Paleoindian. These people had developed a way of life that included hunting the large land animals that flourished during the late Ice Age, though they undoubtedly also hunted smaller animals and gathered plant foods. Presence of these people is

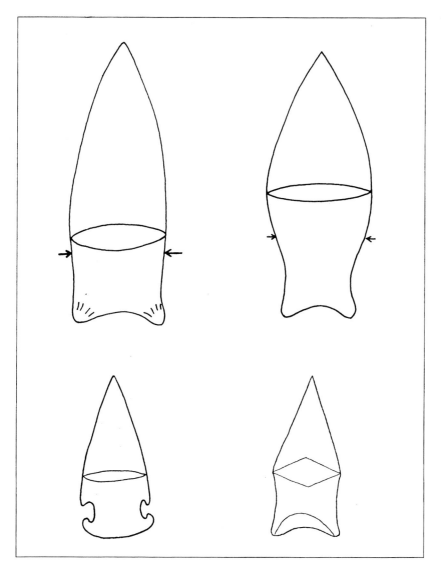

Figure 5. Some Paleoindian Point Types—*top left:* Suwannee; *top right:* Simpson; *bottom left:* Bolen; and *bottom right:* Dalton (Bullen 1975:44, 51, 55–56)

deduced in many cases from their projectile points, which have notches in their sides near the base in addition to the grinding that is the hallmark of all Paleoindian points. This notched style is called Bolen; earlier Paleoindian points discovered in Florida are known as Suwannee and Simpson (Purdy 1981:8, 24, Fig. 11).[3] The origin of these first Floridians presumably was to the north, inasmuch as both the points and the animals have their affinities in that direction. There has been a lot of speculation about southern origins for South Florida's prehistoric cultures, but this relates to more recent periods and will be discussed in succeeding chapters.[4]

So what drew the Paleoindians to South Florida, and especially to the three sinkholes in which human remains have been found? There are several answers.

For one thing, the animals were there. In addition to the giant tortoise, Little Salt Spring produced remains of an extinct large box turtle, the extinct ground sloth, mammoth or mastodon, and an extinct bison, as well as various species that survive today

---

[3]The South Florida finds support the view that the area was not occupied until late in the period. The West Coral Creek points are Bolen (Hazeltine 1983:98–99), whereas a Bolen and the base of a Simpson were found at Myakkahatchee (Luer et al. 1987:146). Cutler Fossil has produced a Bolen, along with one of a less common type Purdy had identified with the late Paleoindian period, the Dalton (Carr 1986:231; Purdy 1981:31–32, Fig. 10t–v). The Harney Flats site, which actually is outside South Florida archaeologically speaking, also appears to be earlier, as both Suwannee and Simpson points are present. It also differs from South Florida in being at the south end of the outcropping of chert on the eastern seaboard, which well may be a factor in explaining why it is larger than sites to the south (Daniel and Wisenbaker 1987:33, iii).

[4]Suwannee points, with their lanceote shape and concave base, bear obvious affinities to the Clovis points identified on the basis of several sites in New Mexico (Spencer et al. 1965:23) and also found in Florida (Milanich and Fairbanks 1980:39). As for the large Ice Age animals found in South Florida, they represent by and large such temperate-zone species as mammoth or mastodon and bison (Clausen et al. 1979:610) and horse (Hrdlicka 1918:30). Considered together, these factors seem to make a conclusive case that the Paleoindians of South Florida came from the north.

(Clausen et al. 1979:610). Warm Mineral Springs has yielded bones from a ground sloth and a saber-toothed cat, as well as such present-day species as panther, deer, opossum, and raccoon (Cockrell and Murphy 1978:6). The inventory from the Cutler Fossil site, in addition to the horse and condor, includes bison, camel, sloth, peccary, jaguar, Florida lion, and dire wolf— so much wolf that it may have been a den. These other finds, however, come from an area for which human association remains highly conjectural (Carr 1986:232).

Also, water was a precious commodity in a land that was drier and cooler than it is today. In fact, Randolph Widmer believes that until 13,000 years ago "the environment of South Florida was too inhospitable for human occupation" due to its aridity (Widmer 1988:189).[5] Sea level, which had fallen as much as 300 feet beneath present levels at the height of the Ice Age, due to the amount of water frozen into the glaciers, still was as much as 100 feet lower than it is today when the first humans entered Florida (Milanich and Fairbanks 1980:37). In such an environment, sinkholes and springs could have been the only sources of fresh water.

The finds in Little Salt Spring and Warm Mineral Springs came from underwater ledges that in Paleoindian times were considerably nearer the surface of the water, if not actually exposed.[6] The authors of the Warm Mineral Springs monograph believe that drinking water was the main attraction of the

[5]Regarding the Ice Age climate, Rhodes W. Fairbridge explains (1984:427) that as the oceans shrink and the climate becomes cooler, there is less evaporation and therefore less rain. "The entire globe gets colder and colder, as well as more and more arid," he says. Locally, "The shores of Ice Age Florida extended out to the edge of the deep waters of the Florida Strait in the east and to the Gulf of Mexico in the west. The present-day peninsula now about 100 km across was then 200 km across."

[6]There is a disagreement as to whether the ledge upon which human remains were found in Warm Mineral Springs was above water when they were deposited. Cockrell and Murphy (1978:1) think it was, and William C. Lazarus (1965:56) says it was as much as 50 feet above sea level 10,000 years ago. Carl J. Clausen and his co-authors disagree, citing (1975:22–24) such factors as the degrees of preservation,

springs (Clausen et al. 1975:31). The dry sinkhole of the Cutler Fossil site also probably functioned as a source of water, albeit in a different manner. Whereas the wet sinkholes were deep enough to extend below the water table, the dry one appears to have cut into an abnormally high deposit of ground water, known in geological terms as a "perched" water table (McGoun 1986:A28).

The fact that sites not associated with sinkholes, such as Myakkahatchee and West Coral Creek, also have not produced any Ice Age animal bones could be due to lack of preservation. On the other hand, it could illustrate that such sites were of lesser importance. Dan Hazeltine (1983:98) cites a large slough near West Coral Creek as the probable source of water during Paleoindian times, but it seems questionable whether the slough would have existed then, given the lower sea level.

The lower sea level of this era also means that our pictures of Paleoindian culture is incomplete. As Milanich and Fairbanks (1980:38) have noted, any coastal or estuarine sites that may exist are located far beneath the sea and beyond the range of existing archaeological technology. Fortunately, these lost sites may not be all that critical for our understanding of the Paleoindian occupation. In the case of Southwest Florida, Widmer argues that the coast would have been little utilized because it would have produced far less food in that period than in historic times. The only brackish estuaries would have been north of Sanibel Island, and these would have been small and also extremely variable in salinity, he says (1988:192).

We can conclude reasonably that the Paleoindians were few in number. Widmer (1988:195) estimates a population of 5,000 on the west coast between the Manatee-Sarasota county line and Cape Sable, and even that estimate seems high. The two springs could not have supported very many people, and the number of

---

the lack of disturbance, the absence of evidence of fire, the deposition of leafy material, and the lack of "drip holes" associated with soft sediments in dry caves. However that may be, the point is that the spring would have been an important resource for Paleoindians in the relative aridity of Ice Age Florida.

46

stone points and tools discovered at West Coral Creek (Hazeltine 1983:98–99) could be attributed to a small group. The same can be said of the assemblage from Cutler Fossil site (Carr 1986:231). In fact, this kind of assemblage is typical of most Paleoindian sites throughout the state (Milanich and Fairbanks 1980:38).

A notable exception lies somewhat outside South Florida, at Harney Flats in northern Hillsborough County, where an area of roughly ten acres was explored. One living area, one activity area, one possible living area, and four possible activity areas were identified. I. Randolph Daniel, Jr., and Michael Wisenbaker (1987:127–128) believe at least two and possibly all three of the site's major sections were occupied simultaneously.

The authors stress that the differentiation of living from activity sites—made on the assumption that retooling would be done in living areas—is significant in that most previous Paleoindian studies, both in Florida and elsewhere, have tended to focus strongly on point types, which means in turn they often are kill sites (Daniel and Wisenbaker 1987:122, 175, 145).

One effect of this emphasis has been to make too much of the idea of Paleoindians as big-game hunters. They were, but they were more. There is no reason to believe they did not make use of small animals, as well as plants. Daniel and Wisenbaker (1987:131) cite finds of grinding stones and of plants within hearths in the Southwest in support of this contention. These finds, however, date from later in prehistory (Purdy, personal communication). Still, large animals make tempting targets in that they provide a lot of meat in return for less effort than would be needed to get the same amount of food from small animals.

Given the size of the Paleoindian groups, they would have had a minimal form of social organization. They undoubtedly had no social grouping beyond that of the band, which Conrad Phillip Kottak defines as a "small group of people—fewer than one hundred—related by ties of actual or believed kinship and/ or marriage." There would have been no division of labor aside from that based on gender or age. Respect would have been accorded to persons with desirable personal qualities—perhaps the aged, due to "their knowledge of ritual as well as of practical matters"—but they would have had no authority over other band members (Kottak 1974:149, 151). One exception to this general rule probably would have been the hunt, especially if the quarry were large. Some sort of discipline would have been necessary, albeit on a temporary basis, if several individuals were to work together efficiently.

Technologically, these first South Floridians utilized a variety of materials in making their tools and weapons. In addition to the stake, Little Salt Spring has yielded a portion of a nonreturning boomerang (Clausen et al. 1979:611). The use of shell is indicated by a spearthrower spur found with a skeleton in Warm Mineral Springs; antler, bone, and teeth also may have been used there, though there is no way to tell now whether the several such objects recovered in the 1950s by William Royal in fact date from Paleoindian times (Cockrell and Murphy 1978:4).[7] The Cutler Fossil site also produced artifacts made from bone (Carr 1986:231).

Generally speaking, however, the people of the Ice Age are best known for their worked stone. It is from stone, for instance, that the projectile points reported for the various sites were fashioned. The West Coral Creek site already has yielded more than 500 stone objects (Hazeltine 1983:99). In addition to the points, they include knives, scrapers, perforators, and spokeshaves, all types that Barbara Purdy (1981:9–26) has reported for the Paleoindian period. Articles of both oolitic limestone and chert are reported for the Cutler Fossil site (Carr 1986:231).

By and large, the objects found seem best suited for hunting animals and preparing them for consumption, thus supporting the assumption that the Paleoindians were primarily hunters. The presence of the mortar, however, shows that they utilized vegetable foods as well. While mortars usually are associated with cultivation, some have been reported for the nonagricul-

---

[7]With some exceptions, the importance of ancient objects cannot be evaluated without context. Because we do not know exactly where Royal obtained his Warm Mineral Springs specimens, we cannot know for sure if they are as old as the human remains. Also, the handling of these objects in years since has raised the possibility of confusion (Clausen et al. 1975:33).

tural Key Marco site (Gilliland 1975:54, 60, Plates 19, 24, 26), where they presumably were used for such gathered foods as nuts.

It is reasonable to assume that whatever clothing these people wore, it was made from animal skins and vegetable fibers. We have not found any of that clothing and possibly never will, given the way such materials deteriorate quickly in a warm moist climate. However, woven vegetable fibers have been identified recently from the 7,000-year-old Windover site near Titusville, where they were preserved in muck (Andrews and Adovasio 1988). Also, two of the bone tools found by Royal and Clark (1960:285) were needles.

As for their shelter, all we can do is speculate. Paleoindians may have used windbreaks of skin or other materials, or utilized the overhangs within the spring sites. In most cases, they probably went without.

Their beliefs also are a matter of speculation, though we can reasonably assume they felt there were spirits or forces present in nature with which they needed to reach some sort of accommodation, if they were to survive. The form of that accommodation is in question. Frederick Webb Hodge (1910:368), referring to North America in general, said, "The Indian is not satisfied with the attempt to avoid the ill will of the powers, but he tries also to make them subservient to his own needs." Barbara Purdy (personal communication) disagrees: "Everything that I have read emphasizes that the Indians did *not* attempt to make the spirits subservient but instead they always felt subservient to the spirits." Cockrell and Murphy (1978:1) interpret a skeleton in Warm Mineral Springs as a burial covered with stones, thus implying belief in some sort of soul or qualities that transcend life. Clausen and his co-authors (1975:32) assert, however, that the human remains belong to persons who fell into the spring and either drowned or died of exposure.

From findings at Bay West, a site just south of the Lee County line in Collier County, John Beriault and others (1981:55) see the existence of a water mortuary tradition in later years and speculate that water could act as a barrier between living persons and ghosts or supernatural beings. Such a tradition could extend backward in time into the Paleoindian period. On a more prosaic level, water burial would be a good means of keeping scavenging animals away from corpses. In fact, some of the Bay West bones have been gnawed (Purdy, personal communication).

All in all, the Paleoindians had developed a way of life that would have continued to serve them well had conditions remained the same. Unfortunately, conditions were not remaining the same. The large animals were on their way out.

The pivotal factor seems to have been a more moist climate. This change allowed the Paleoindians to increase in number and thereby contribute to their own demise by putting greater pressure on resources. By 8,500 years ago many of the animals that thrived during the Ice Age had become extinct (Milanich and Fairbanks 1980:42). The extent to which hunting caused the demise of the large animals is a matter of conjecture. Donald K. Grayson (1977:692) argues that the broad range of extinctions traced to that period, encompassing as many as thirty-two genera of mammals and ten of birds, is too great to be laid at the feet of *Homo sapiens*. Still, pressure from human beings must have been a factor.

The result was emergence of a way of life we call Archaic, the subject of the following chapter. It is hard to say whether the Archaic inhabitants of South Florida were descendants of the Paleoindians or whether they supplanted them. The nearest site to South Florida exhibiting any evidence of the transition is Republic Groves in Hardee County. While most of the items found in a swampy area there belong to the Archaic period, other finds include a Waller knife and a Dalton-like core from which points would have been chipped (Wharton et al. 1981:77). Both Waller and Dalton are identified by Purdy (1981:24, 31–32) as belonging to the late Paleoindian period.

The best evidence of the transition, however, comes from the Nalcrest site, east of Lake Wales in Polk County. Once again the finds were made underwater, around the edges of a lake. They consist of an extraordinarily large number of small chipped stone tools, generally worked on only one side. Points of the Bolen series, the same style of Paleoindian projectile point

50

found at West Coral Creek, are present, and so are points else-where identified with Archaic times (Bullen and Beilman 1973:3).[8] Milanich and Fairbanks speculate (1980:46) that the site may have been used for the processing of a single resource, perhaps cane or grasses.

Even here, however, we cannot be certain whether these people were the sons of the pioneers or whether they had displaced them.

[8]Projectile points reported from the Nalcrest site include Clovis-like (frag-ments), Suwannee, Santa Fe, Beaver Lake, Dalton-like, Bolen Plain, and Bolen Beveled from the Paleoindian period, plus Archaic Stemmed, Culbreath, Citrus, Hernando, Pinellas, and miscellaneous side-notched from the Archaic (Bullen and Beilman 1973:3).

THREE

Living Off the Land
The Enduring Hunting and Gathering Societies

With the demise of the Paleoindian system, there arose all across North America an opportunistic and long-lasting way of life that archaeologists call Archaic. In South Florida, it was even more long-lasting than elsewhere.

To Peter Farb, Archaic people, unlike their big-game hunting predecessors, "specialized in nothing, but they were versatile in attempting everything. . . . In a remarkable diversity of environments . . . [they] invented fish spears, snares for trapping rodents and birds, darts for bringing down small game, baskets for collecting roots, stones for grinding seeds" (Farb 1968:206). Evidence of this change is an oak mortar at Little Salt Spring that dates from roughly 7000 B.C., the beginning of the Archaic period (Clausen et al. 1979:611).

In other words, they ate whatever was available, making the tools and weapons necessary from whatever was available. In much of Florida, as Jerald T. Milanich and Charles H. Fairbanks noted (1980:19), this meant "the use of shellfish, either from the freshwater streams and lakes or from the coastal lagoons," though "the evidence of their tools and weapons suggests that hunting, especially of deer, was still a major concern."

This opportunism was necessary, Milanich and Fairbanks explain (1980:50), because the climate was becoming drier, rising sea level was reducing the amount of land available and some animal species had become extinct. The result, they say, probably was "increased use of certain old food sources and first use of new ones. We can surmise that acorns and hardwood nuts, because they were abundant and accessible, were one old source that became more important, at least on a seasonal basis. Freshwater snails appear to be first exploited along the inland rivers at this time and, perhaps, oysters were harvested in large numbers along the Gulf Coast."

Conventionally, the Archaic is defined in time as the period between the Paleoindians and the rise of regional specializations marked by the introduction of pottery and/or agriculture and mound-building; Robert F. Spencer and Jesse D. Jennings (1965:17, 57) use different criteria at different points. Willey and Phillips (1958:202) say the Archaic generally gives way to sedentary agricultural cultures, but that definition is not especially useful in South Florida because there were relatively sedentary post-Archaic cultures, in terms of time, that were nonetheless basically Archaic in terms of subsistence.[1]

Any attempt to speak of an Archaic period in South Florida is fraught with a major difficulty: Where do you put the end point? The Archaic stage of culture, a nonagricultural lifestyle based on hunting, collecting, and fishing, persisted into the contact period in much of South Florida. Chronologies and taxonomic schemes applied to northern Florida and elsewhere in the Southeast are not necessarily applicable to South Florida. The use of the word *Archaic* as a period in time rather than a lifestyle leads to much confusion, but it is unavoidable given its ubiquity in the literature.

At the time of Spanish contact only the Calusa, with their intensive fishing and shellfish collecting, had the sort of sedentary existence that could make them distinct from the Archaic way of life. Other South Florida groups, while they could stay put for long periods, were driven into nomadism at least part of the

[1]The only comprehensive look at prehistoric South Florida is an unpublished manuscript written by Goggin in the late 1940s, in which he implicitly made mound-building his marker for the end of the Archaic period (m.s.:711). That makes sense, and mound-building might have been useful to determine the entire Glades period had the practice persisted into historic times. But, there is no evidence that historic groups built the mounds they used, unless the accumulation of refuse can be called "mound building." There are four mounds on the western end of Horr's Island in Collier County that appear to have been built no earlier than late prehistory, but McMichael notes (1982:100) that "No artifacts were recovered from any of the mounds that would aid in determining their age or origin." One mound on the eastern end of the island was certainly used in the historic period. Glass beads were found by Stirling (1931:171) in his 1930 excavations.

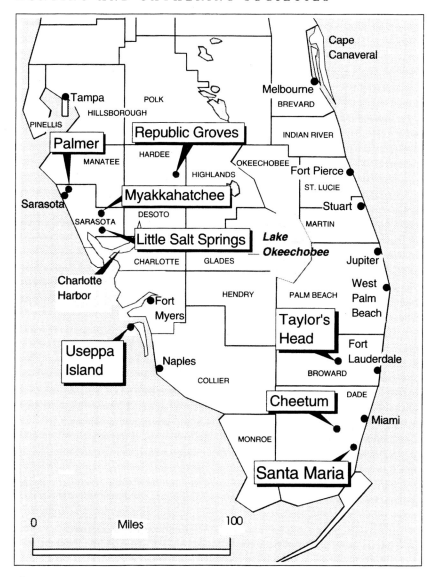

Figure 6. Major Sites of the Archaic Period

year in search of food. Milanich and Fairbanks (1980:20) take note of the problem in saying, "There is some evidence to suggest that the extreme southern third of Florida (outside of the Lake Okeechobee Basin) remained in what was basically an Archaic stage until the coming of the Spanish."

For most of Florida, Milanich and Fairbanks (1980:20, 23) use a 6500–1200 B.C. Archaic time frame divided into Early (6500–5000 B.C.), Middle (5000–3000 B.C.), and Late (3000–1200 B.C.) Archaic subperiods, followed by a 700–year transitional period marked by the rise of specializations probably including some agriculture. Robert S. Carr and John G. Beriault (1984:1–2) follow a similar scheme except that they include the period until 500 B.C. in the Archaic.

Randolph J. Widmer (1988:64–73) has a somewhat different chronology divided into Early Archaic (7900–5000 B.C.), Middle Archaic (5000–2000 B.C.), and Pre-Glades (2000–700 B.C.), this last a terminology originated by W. A. Cockrell (1970:55) for what others call the Transitional period.

I prefer a system of Early Archaic (7000–5000 B.C.), Middle Archaic (5000–3000 B.C.), Late Archaic (3000–1500 B.C.), and Transitional (1500–500 B.C.). Widmer himself notes (1988:63) that the Paleoindian pattern persists in South Florida until 7000 B.C., and his compression of Late Archaic and Transitional into a Pre-Glades period seems to be based on reinterpreting data from the Palmer site, Sarasota County, in a manner I cannot accept.[2] As for the term *Pre-Glades*, I feel that *Transitional* better describes the nature of the period.

In South Florida, the researcher must once again, as in the

[2]In deciding upon a 2000 B.C. beginning for the Pre-Glades period, Widmer apparently has chosen to put the advent of fiber-tempered pottery at the Palmer site at 2000 B.C., based on the fact that a single sherd was found one foot above the level from which two radiocarbon dates in the 2000 B.C. range were obtained. That conclusion to me is unjustified, especially inasmuch as all other pottery at the site comes from near or above the level where a date in the 1500 B.C. range was obtained (Widmer 1988:68; Bullen and Bullen 1976:13, Table 2). It is dangerous to draw conclusions from a single sherd, or even a handful of sherds. Unfortunately, this kind of conclusion has been made more than once, especially in classifying small sites.

Paleoindian period, make his inferences from a limited amount of data. For the entire Early Archaic there is not a single definite site, and there are only four with good time controls for the Middle Archaic. In fact, it is the presence of sites that separates the two periods.

Perhaps this scarcity of sites is indicative of the fact that times were rough, which in turn would encourage the development of broad-based, opportunistic subsistence strategies. Or is it the result of incomplete research? Widmer accepts the former view. The "pollen profile . . . suggests an extremely arid environment in south Florida," he says by way of explanation (1988:202). Under such conditions, he says, the animals that survived into later periods—along with the humans—could have retreated into North Florida while those species that could not retreat, such as the sloth, died out.

"Only scattered sinkholes and cenotes would have been found in the coastal zone," Widmer says (1988:202). "The carrying capacity of south Florida as a whole would have been lower, since well-watered areas became scarcer, and thus populations would have been smaller. The arid interior zone . . . would be utilized to a much lesser extent than in the earlier Paleoindian period." Purdy (personal communication) says the freshwater snails of which Milanich and Fairbanks speak were not present in the early Archaic.

Widmer (1988:203) does not believe the depopulation was total, and there are various reasons why the lack of sites does not mean a lack of population. To begin with, much of South Florida, while it may or may not have been attractive to prehistoric opportunists, is quite unattractive to twentieth-century archaeologists. The terrain is difficult, and the combination of heat and mosquitoes makes fieldwork especially unpleasant in the summer, which is the only time of year that many archaeologists can get away from their teaching duties. Aerial detection of sites, such as that used in the National Park Service surveys of Everglades National Park and the Big Cypress National Preserve, easily could fail to spot small Archaic campsites, because they presumably would be buried at some depth due to the amount of time since their occupation.

Another factor is site destruction because of natural processes. John E. Ehrenhard (1982:22) says Archaic occupation was most likely "in the fringe areas along the east and west coasts," and Robert Taylor (personal communication) notes that some exposed coastal sites in Everglades National Park were eroded.

Again, it is important to remember that the term "coast" is a relative one. As was true of Paleoindian times, most of what was coast then is underwater today, and sites there are inaccessible. A steady rise in sea level during the last 6,500 to 7,000 years has been postulated by David W. Scholl, Frank C. Craighead, Sr., and Minze Stuiver (1969:562–564).

In more accessible areas, it may be significant that the older dates are due generally to the newer research. For instance, Dade County avocational archaeologist Dan Laxson, who excavated extensively in the Hialeah area in the 1950s and 1960s, reports on various occasions having encountered limestone bedrock at depths ranging from six to thirty inches (Laxson 1957:18; Laxson 1964:177). Such "bedrock" is in fact a calcareous lens of fairly recent origin, beneath which evidence of occupation has been found (Mowers 1972:129–131). Mowers believes these lenses are natural; Palmer and Williams (1977:25) say that is probably true in most cases but that in many instances aboriginal fires released from shell calcium compounds that facilitated lens formation.

Finally, it is very difficult to detect a preceramic site, inasmuch as ceramics often are the only evidence a location was occupied in prehistoric times. In fact, most Archaic sites that have been found date from the Transitional period and are marked by pottery in which both fibers and sand were used as tempering agents.[3]

[3]Clay is the basic material of pottery, but many clays will crack or become distorted during firing unless they are mixed with a nonplastic "tempering" substance. Sand is the most common such substance in prehistoric Florida pottery, though fiber and limestone also have been used. The supposedly untempered St. Johns ware found in South Florida was in fact made from a clay that was naturally mixed with fossil freshwater sponge spicules (Mowers 1975:8; Milanich and Fairbanks 1980:233).

To sum up, sites throughout South Florida, at least on the coast, *could* have been occupied throughout the Archaic period. The general tendency of populations to expand in both numbers and range over time and the opportunism of the Archaic people argue that many sites as yet undetected, and quite possibly undetectable, in fact *were* occupied.

To find the first hard evidence of Archaic occupation, we must return to that area near the west coast that has provided most of what we know about South Florida's Paleoindians. The earliest presence seems to have been at four sites that also show evidence of Paleoindian occupation: Little Salt Spring in Sarasota County, Bay West in Collier County, Republic Groves in Hardee County, and Myakkahatchee in Sarasota County.

Carl J. Clausen and his colleagues, in addition to reporting on the mortar, believe, on the basis of a small sample, that a muck slough adjacent to Little Salt Spring holds possibly 1,000 burials (Clausen et al. 1979:609). Dates of 4000 B.C. or earlier have been obtained both by the radiocarbon method and through a distinctive point type identified by Ripley P. Bullen as the Newnan (Bullen 1975:31; Clausen et al. 1979:612). At Bay West, Beriault and his colleagues (1981:50) report discovery of thirty-five to forty skeletons from peat deposits in a pond, in contexts suggesting dates of 4400 to 4900 B.C. Republic Groves has yielded remains of at least thirty-seven persons in muck along a freshly cut canal bank, much in the manner that the Vero finds came to light. Here, artifacts and radiocarbon dates suggest occupation dating from 4600 B.C. (Wharton et al. 1981:56, 76, 78). Myakkahatchee produced no skeletal material, but did yield at least fifty stemmed projectile points or knives, plus some reworked tools. Among the types identified are Putnam, Levy, Marion, and Newnan; all are identified as Middle Archaic by Purdy (Luer et al. 1987:146; Purdy 1981:Fig. 15a–d).

What makes the period after 5000 B.C. more conducive to settlement, Widmer says in synthesizing the work of several others, is a higher water table and more surface water in the interior of northern South Florida, though the interior to the south remains arid. He says the rising water table led to creation of the slough in which the Little Salt Spring burials were found

59

and cites the smaller number of burials at Bay West as illustrating that the latter site had fewer inhabitants because it lay in a less desirable, more arid zone (Widmer 1988:204).

The Beriault group suggests the possibility of a tradition of burials in moist peat beneath the water level of springs, ponds, or sloughs. They hypothesize that the wooden stakes found in association with burials at Little Salt Spring, Bay West, and Republic Groves were used to hold burial biers or mats in place. As noted previously, they also say water burial may have been a means of erecting a barrier to keep ghosts and supernatural beings away from the world of the living (Beriault et al. 1981:54–55).

The Beriault group also comes up with yet another reason why the number of Archaic sites may be underreported. Cultural resource inventories do not include intensive searches for such features as ponds and sinkholes, they say (1981:55). Ehrenhard (1982:15) agrees with this point as regards Everglades National Park and the Big Cypress.

So who were these people of the water? Their opportunism is made evident by the wide range of tools and weapons reported. The inventory from Little Salt Spring and Bay West includes wood that seems to have been shaped into tool handles or tools, twigs and leafy matter that may have formed biers, deer-bone projectile points and atlatl hooks, antler and stemmed stone points (Clausen et al. 1979:612; Beriault et al. 1981:45–49). Republic Groves was especially rich in such objects. Besides the wooden stakes, these artifacts include a number of Newnan and Florida Archaic Stemmed (a class that includes Putnam, Levy, and Marion) stone projectile points; other stone fashioned into drills, scrapers, knives, choppers, a combination grinder and anvil, a bone, and beads; antler worked to produce an atlatl hook, a plummet, and several ornaments; bone possibly used as awls, a flesher, pins, and knives; and three shark's teeth (Wharton et al. 1981:64–66).

Interestingly, of the two sites closest to the coast, shell and shark teeth are reported for Little Salt Spring but not for Bay West (Clausen et al. 1979:612; Beriault et al. 1981:49). This finding may indicate nothing more than the fact that Bay West

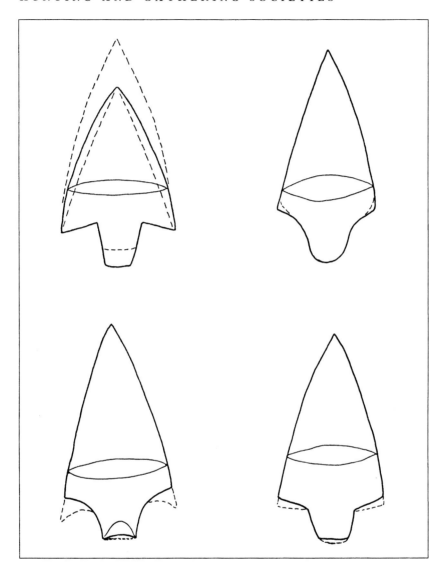

Figure 7. Some Archaic Point Types—*top left:* Newnan; *top right:* Putnam; *bottom left:* Levy; *bottom right:* Marion (Bullen 1975:31–32)

was excavated under less than ideal conditions. On the other hand, it may suggest that the Little Salt Spring people had more access to the coast, either through travel or through trade. This supposition is not all that surprising; access would have been easier down the Myakka River from Little Salt Spring than over the Gulf Coast sand ridge from Bay West.

More evidence along this line is the discovery of shark's teeth at Republic Groves, which is roughly fifty miles inland even today but lies near the Peace River. This discovery also suggests the possibility that the Archaic people had canoes, a not unreasonable assumption in light of the fact that a canoe dating to 650–1000 B.C. has been found northwest of Lakeland, not far from the headwaters of the Peace (Bullen and Brooks 1967:102–105). Florida sites in fact have produced nearly 200 canoes, the oldest from DeLeon Springs, several miles north of DeLand near the St. Johns River, dating to 3000 B.C. (Newsom and Purdy 1990: 164–180).

As noted before, occupation on the coast itself may have been obliterated by rising sea level. In addition, Cockrell (1970:88) suspects that Archaic people had not developed the sort of specialized subsistence system that would have allowed extensive exploitation of coastal resources and perhaps had maintained only seasonal camps there. Dana Ste. Claire disagrees, at least as regards Northeast Florida; she says (1990:189), "Recent evidence suggests that areas near the coast were utilized as early as 6000 B.C."

In addition to being mobile, as almost all nonagricultural people must be in the absence of the sort of concentrated food resources that in practice exist only in a marine setting, these early South Floridians undoubtedly had some contacts with people outside the area. A pendant of serpentine-like stone available no nearer than the southern Piedmont was found at Republic Groves (Wharton et al. 1981:65). A bead possibly of quartzite was reported from Bay West (Beriault et al. 1981:49), but Purdy (personal communication) says it could have come from a local phosphate deposit. In neither case is the age of the object certain.

Adding to the intriguing nature of the mortuary hypothesis is the possibility that, in addition to dating back into Paleoindian times (assuming the Warm Mineral Springs remains were buried underwater), it may date forward into the historic period. Little Salt Spring has yielded part of a wood tablet similar to those reported for the protohistoric component of the Key Marco site (Clausen et al. 1979:612, Fig. 4; Cushing 1973:426, Plate XXXIV; Gilliland 1975:75–80, Plates 31–38), and Bay West has produced wood fragments that could be the same (Beriault et al. 1981:45–46). In other words, what we may have is a religious, or at least mortuary, tradition that spans 10,000 years and at least two cultural changes. This possibility is not all that far-fetched when you consider the cultural differences between Biblical Israel and today's Israel, for instance.

The fact that with one exception these sites are found only in a portion of the Gulf Coast does not preclude their presence elsewhere, given the problems of detecting them noted earlier in the chapter and the unsuitability of the sandier east coast for preservation of bone and wood. That exception is the Gauthier site in mid Brevard County, where the projectile points associated with burials (notably Culbreath and Newnan) suggest dates as early as 4000 B.C. in the Middle Archaic. The site is on a sand ridge at the point where the St. Johns River comes fairly close to Indian River, making it a good location for water travel (Jones 1981:81, 83).

There is the Windover site near Titusville, burials from which have been dated to 6000 B.C. in the Early Archaic (Doran and Dickel 1988:365), but it is arguable whether Windover can be considered a South Florida site. The thirty or so miles separating Windover from Gauthier is enough to put Windover outside South Florida as defined archaeologically in later periods and the distinction could hold true in the Archaic as well.

As for Archaic social organization, nothing found thus far suggests any differentiation beyond age and sex roles and deference to skill in a given task, as in the Paleoindian pattern. Milanich and Fairbanks (1980:50) say "social groups would come together to form larger populations at certain times of the year and break up into small family groups at other times, according to the resources being collected or hunted and the need for sharing or nonsharing." They refer to areas farther north in the state, but their conclusions could apply to South Florida as well.

Statements about plant food must be speculative in the absence of evidence, but there is no reason to believe these people were any less opportunistic here than they were with regard to animals, as reflected in the range of bones found at their sites. Frederick Webb Hodge, speaking of the Americas north of Mexico in general, notes (1911:466) that "no pure hunter stage can be found, if it ever existed."

Given the apparent presence of matting, it is logical to assume these people had baskets in which to carry and store plant foods. A range of grasses and related plants have been identified from Little Salt Spring (Brown and Cohen 1985:24), but it is not clear how much those species were eaten by humans. Further, the mats that held the dead in burial probably held the living in sleep. Either skins or woven plant fibers could have been used for clothing. As for shelters, if there were any they probably were thatch huts that could have been similar to the chickees of the historic Seminoles.

All of this discussion is a long way of saying Paleoindian and Archaic people were a lot alike. Those differences that did exist can be explained in terms of environment. Each lived in small groups, ate what was available, made tools and weapons from what was available, and sought to stay on the good side of the spirits.

One thing that seems certain is that the Archaic diet was adequate. Lorraine P. Saunders has analyzed the Republic Groves remains and has concluded (1972:17–39) that the people represented were robust and strongly muscled. Estimates of their height range from 5 feet 2 inches to 5 feet 8 inches—not tall by twentieth-century standards but not that short either. Pathologies were minor—a suggestion of hookworm and some arthritis being most notable—while the teeth showed a lot of wear but few cavities, a typical pattern among people who eat a gritty diet but no refined sugar.

After these sites there is another gap in the record. For the period between 4000 and 3000 B.C., the hard evidence consists of a single radiocarbon date (3675 B.C.) from a midden on Useppa Island (Milanich et al. 1984:270). Further, this date was obtained from shell and such dates are controversial. Sears argues (personal communication) that such dates are influenced by the calcium carbonate content of the water in which the shellfish lived.

Then comes a cultural explosion, at least in comparison with what went before. Another Useppa Island midden date is in the 3000 B.C. range, and so is one for Marco Island that Widmer says is considered to be accurate (Milanich et al. 1984:270; Widmer 1988:71). Claudine Payne (1992:1) sees year-round occupation on Horr's Island by 2800 B.C. Also, for the first time we have solid evidence for Archaic occupation in Southeast Florida. There are dates of roughly 3000 B.C. for the Cheetum and Santa Maria sites in Dade County and Taylor's Head in Broward. However, there is some question about the context of the charcoal from which the Santa Maria date was obtained (Newman 1986:55; Carr et al. 1984:174, Table 1; Masson et al. 1988:346).

This expansion is not especially surprising, given the significant environmental changes associated with this time. Somewhere between 3500 and 2500 B.C., Widmer says, in again synthesizing the work of others, that "dramatic changes occurred in the environment of south Florida. . . . The shift to a modern floral *composition* probably occurred. . . . The water table had risen to a position high enough to prompt the beginning of the hydric regime in the interior of south Florida, notably the appearance of Lake Okeechobee" (Widmer 1988:206–207). Brown and Cohen note (1985:21) that the Little Salt Spring area became more moist about that time. Once this transition was complete, Florida had reached essentially its contemporary environmental condition. The combination of increased drainage from the wetter interior and the decrease in sea-level rise led, Widmer says (1988:207), to the formation of "brackish estuaries . . . mangrove forests, tropical marine meadows, and coastal marshes"—in other words, the rich coastal ecosystem that would be utilized by the historic Calusa.

Alan McMichael has reported on a massive oyster shell midden on Horr's Island, near Key Marco in Collier County, that

dates to 2000 B.C., and he believes the area probably was popu-
lated prior to that date (1982:46–59, 81). A similar inference is
made for the Palmer site (Bullen and Bullen 1976:49). In neither
case is there any hard evidence. An oyster and clam midden at
Useppa Island contained a burial dated to 2160 B.C. (Marquardt
1987c:4).

Another group of Southeast Florida radiocarbon dates are
bunched at roughly 2000 B.C. Four come from Dade County
sites, two tree islands in the west and a midden and cemetery on
Biscayne Bay south of Miami River (Carr 1979:10, 1981:4).
Wesley Coleman has reported (personal communication), but
not published, a similar date for a site in western Broward
County.

Given the attractiveness of this new environment, it is strange
indeed that after 2000 B.C. there appears to be another 500–year
gap in the archaeological record. Milanich (personal communication) thinks this gap is illusory, and it could easily be explained by the difficulty of finding late nonceramic sites. Signifi-
cantly, the one site from that period—Palmer—also is one of the
earliest ceramic sites in South Florida.

Fiber-tempered pottery definitely was on hand by 1500 B.C.,
both at Palmer and at Caxambas Point on Marco Island (Bullen
and Bullen 1976:13, Table 2; Cockrell 1970:60). It also is known
in stratigraphic context at Useppa Island; Marquardt (personal
communication) believes this pottery could date as early as 1500
B.C. Within another 500 years, by roughly 1000 B.C., there are
numerous sites marked by the presence of pottery tempered
with fiber or fiber and sand. Such sites are relatively common in
Sarasota, Charlotte, and Brevard counties and are found elsewhere as well (Site cards, Bureau of Historic Preservation,
Florida Department of State, Tallahassee).

Could the 500-year gap preceding the first ceramic sites be due
to a lack of continuity that reflects the arrival of a new people,
moving into a land vacated by the earlier inhabitants for reasons unknown? If so, did these newcomers arrive from the north
or from the south? One argument against the arrival of a new
population is the persistence of a generalized subsistence pattern and of artifact types, though this situation also could be

explained by the constraints of the environment. A better argument is the apparent persistence of the burial tradition. Amy Felmley (1990:265) believes that the uniformity of burials throughout the region "suggests continued interbreeding between culturally distinct populations" from the Late Archaic through European contact. Still, the question remains open in the absence of further research. If there were intruders, however, they almost certainly came from the north.

There have been over the years a number of theories about migration from the south, some of them fanciful.[4] The most credible scenario is that put forth by William H. Sears. He argues that as early as 2000 B.C. small groups of people from the south had entered Florida through the Everglades and migrated northward through the middle of the state. Among other things, he believes these people were responsible for the introduction of the first Florida pottery, fiber-tempered ware that makes its appearance about 2000 B.C. in the St. Johns River valley and even earlier in the Savannah, Georgia, area (Sears 1977:3, 5; Ford 1969:12).[5] Additional evidence in support of that view comes from Robert W. Long (1984:118), who has pointed out that 61 percent of South Florida's plant species come from the south,

[4]Fanciful writings on possible southern origins for South Florida cultures generally were written some years back by non-archaeologists. Notable among such writings are George A. Hewitt's claim (1898:137) of Aztec origins for people of the upper Keys, the idea of Mayan origin for a Key Largo site as put forth by John C. Gifford and Alfred H. Gilbert (1932:313), and Doris Stone's contention (1939:218) that the use of shell tools diffused from the Bahamas and the West Indies. Regarding the Stone idea, environmental similarities make independent invention a more logical explanation.

[5]The thoughtful earlier analyses, aside from that of Sears, discount southern origins. Hrdlicka (1922:114–116, 127, 130–131) says the head shape is wrong. John W. Griffin (1943:89) cites the lack of manioc cultivation, and Rouse (1940:63) notes that the Melliac, the type of Antillean pottery found closest to Florida, is not found in Florida. Sturtevant concludes (1960:43) "that the supposed ethnological parallels between the Southeast and the Antilles are either nonexistent or can be explained on grounds other than diffusion across the Straits of Florida."

and Hale G. Smith, who has argued (1951:241) that *Zamia* species from which root starches are obtained were brought to Florida from the south by Indians.

Still, the case is weak. Even allowing for problems in detecting Archaic sites there, the negative results from surveys of Everglades National Park and the Big Cypress argue against the Sears theory (McGoun 1985). As for the plants, they need not have been brought north by humans. And then there is the fact that fiber-tempered pottery does not show up in South Florida for as much as 500 years after it makes its appearance in the St. Johns River valley (Milanich and Fairbanks 1980:60–61; Bullen and Bullen 1976:49).

The fiber-tempered pottery of the Georgia coast and the St. John's valley is the earliest pottery found in what today is the United States. In fact, recent research at Lake Monroe on the St. Johns River has indicated the earliest pottery there predates 3000 B.C. (Purdy, personal communication). Bullen (1961:106) sees the North American fiber-tempered pottery as an indigenous development, while Sears (1977:12) envisions a series of small-group movements from the south and does not see the lack of fiber-tempered pottery in intermediate locations as a refutation.

Regarding the possibility that the sand-and-fiber-tempered ware found in the Lake Okeechobee Basin is an outgrowth of the ware from the St. Johns valley, Sears (1977:7) says, "It lacks decoration and the shape is wrong." Elsewhere, he says (m.s.:6) the open-bowl tradition of Florida "is really paralleled only in the Caribbean and in some parts of lowland South America."

The most telling argument against southern origin, however, is that the objects we have found invariably show affinities to the north. This point has been made at various times for various areas by A. L. Kroeber (1953:67), M. W. Stirling (1936:356; 1940:120–121), and John W. Goggin (1944:31). Gordon R. Willey (1949:129–130) sees the later Southeast Florida cultures as based firmly on the Archaic of the eastern United States, while Clausen and his co-authors (1979:612) see them as "attributable to compression of an Archaic culture into the southern Florida area by later, possibly agricultural, peoples entering the peninsula from the north."

Whatever its origin, the appearance of the first pottery would be the most significant change in South Florida culture in millennia. Coincident with its arrival were major cultural changes on the southwest coast and in the Lake Okeechobee Basin. In the latter case, the southern influence appears to be inescapable.

FOUR

Earthworks and Effigies

Hopewellian-Related Societies Around the Big Lake

The flames engulfed the north end of the platform, eating through the pine posts that supported it and charring the bird and animal figures. Once the structure was sufficiently weakened the platform collapsed, sending the bones tumbling into the pond below. The fire happened some time in the early centuries of the Christian era, about the time Rome was falling to invaders from northern Europe and the classic Maya civilization was arising in Mesoamerica. Whether the fire was accidental or deliberate, it undoubtedly was an important event in the history of a complex but little-understood culture that had arisen some centuries earlier in the Lake Okeechobee Basin.

Did this culture evolve out of the Archaic way of life that had persisted in the area for millennia, or was it introduced from elsewhere? If it was introduced, are its affinities to the Midwest, site of the Adena and Hopewell complexes, or to Mesoamerica or South America? Most of what we know about the specifics of this phenomenon come from the work of William H. Sears at Fort Center, a prehistoric site on the west side of Lake Okeechobee named for the stockade built there during the Second Seminole War of 1835–42. Sears's answer to these questions would be "Yes."

Sears, who has postulated (1977:11) that many prehistoric cultural traits reached North America from the south by way of Florida's watery interior, believes (1982:191) the Fort Center site was occupied possibly as early as 1000 B.C. by persons already in the area. These people lived along the banks of Fisheating Creek "and on small house mounds well out in the creek meander belt" and, like their predecessors, had an egalitarian society (Sears 1982:192).

At some point during the next 500 years, however, they obtained one important thing that no previous Florida society had possessed: horticulture, specifically the cultivation of

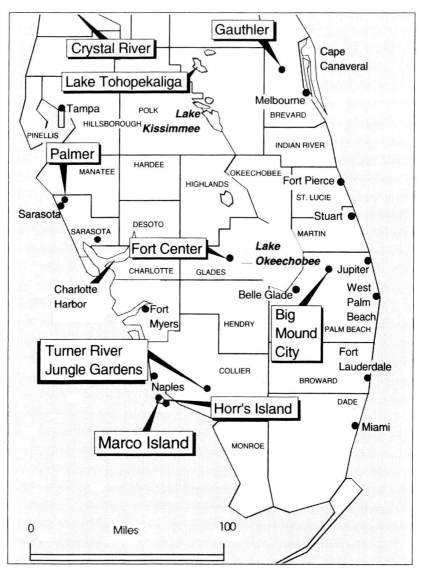

Figure 8. Major Sites, 500 B.C. to A.D. 800

maize. Three overlapping circular ditches that Sears believes were built to drain farm plots were excavated early in the site's history, and maize pollen was found both within the last and largest of these ditches, which Sears calls the Great Circle, and in midden deposits along the creek bank at the ends of that ditch. Radiocarbon evidence indicates that the Great Circle was complete by approximately 450 B.C. (Sears 1982:178, 193). This is roughly the same time that sand-tempered pottery shows up at the site (Sears 1982:Figure 7.1), which suggests the possibility that the maize farmers were newcomers, though not everyone accepts Sears's findings without reservation.[1] Sears (1982:186) rejects the idea that these ditches could have been ceremonial in favor of the explanation that they were cut through the impermeable hardpan stratum that underlies the site in order to allow drainage.

While Fort Center is the only site for which good data are available, it looks as if these people may not have been the only ones digging circular ditches then. Robert S. Carr has located seven other sites containing similar features. Two are in the Fisheating Creek area, one to the north in Glades County, one on the Caloosahatchee River, one south of Lake Okeechobee in Hendry County, and two near Miami (Carr 1985:290).

Stephen Hale is even more expansive. He says (1984:180) that

---

[1]The presence of maize at Fort Center remains controversial nearly a decade after publication of the Sears book on the site. Carr (1985:300) remains skeptical, while Bruce Smith (1986:38) questions the association of maize with the circles marking the first period of occupation. William Keegan (1987:339) says the timing of introduction and cultivation remains an open question, though he does accept the presence of maize at some point. William Marquardt (personal communication) points to the fact that only maize pollen, and not maize itself, was found, while Barbara Purdy (personal communication) cites the absence of cobs. William Johnson (1990:212) wonders if the Fort Center pollen might date from historic times, and Morton Kessel (1991:94) sees the weight of evidence being against horticulture. I remain in Sears's camp. While it can be argued that the finds of pollen in association with certain earthworks can be the result of later contamination, the presence of pollen in the pigment on one of the carved wooden birds and in coprolites (Sears 1982:18, 120) is not so easily explained away.

complexes "with sequences of construction and architectural style almost identical to those at Fort Center" are found on all sides of Lake Okeechobee, ranging as far north as Lake Tohopekaliga and east to the Big Mound City area in Palm Beach County, as well as south into Hendry County and west into Glades. He refers to the entire Fort Center complex, not just the circular ditches.

"The basic architectural plan," Hale says (1984:180, 183), "consisted of a midden surrounded by a large semicircular sand ridge. Extending from the semicircular sand ridge are two parallel linear sand ridges terminating at a large sand mound, which is usually larger than the others at each site. This principal sand mound is surrounded by a small semicircular sand ridge. Shorter sand ridges bordered by shallow depressions or paired linear ridges also extend from the main semicircle. . . . The linear ridges are aligned parallel to the surface water flow to minimize erosion and the labor required for maintenance."

Sears (1982:193) suggests that horticulture may have been introduced into Florida directly or indirectly from Colombia. William Denevan (1970:647–654) reports widespread use in South America of drained-field cultivation techniques, including "ditched fields, mainly for subsoil drainage," though none of the formations he cites is strictly analogous to those found at Fort Center. Sears believes further that the system was in turn exported to the Midwest, where it underwent a fundamental change. Whereas the circles performed a necessary drainage function at Fort Center, "in a wet environment with frequent flooding and a groundwater table only 24 to 30 inches down," in the Midwest they "became ceremonial and developed other forms and variations" (Sears 1982:193–194).

Sears stresses that the northward migrations into Florida were "small-group movements." Therefore, the early maize cultivators—he estimates no more than one or two families at a time in this period—could have been either newcomers from the south or members of a local culture that absorbed new ideas (Sears 1977:12, 1982:193). As for social organization, Sears notes (1982:193) that people living on the site could have built at least the two smaller ditches and possibly the largest one as well.

Denevan (1970:653) says most of the drainage techniques he considers "could have been constructed and managed through small-scale family and community cooperation."

Unlike Sears's theory about earlier migrations from the south, a theory with which most other anthropologists disagree, his views here have met with widespread, though not universal, acceptance. There is the corn itself, at a date "very early indeed for the eastern United States." And then there is what Sears interprets as "large quantities of shell . . . burned into lime" that by all indications "was used in the preparation of dry stored corn, in the Mesoamerican technique." Sears notes that the maize "could have come from anywhere between Colombia and Veracruz" but leans toward a South American origin on the basis of the ditches (Sears 1977:7–9).

Carr says the presence of circular ditches in Dade County, provided they can be dated earlier than those ditches in Lake Okeechobee Basin, could give credence to Sears's hypothesis of movements from the south. Carr, however, remains skeptical of the migration theory (Carr 1985:30).

Thomas J. Riley (1987:296) supports the idea that maize diffused from the south through Florida and on northward. Interestingly, he based his views on incorrect Fort Center data.[2] Donald W. Lathrap (1987:349–360) also supports the Sears concept, saying it is certain that one strain of tobacco entered the present-day United States by sea. William F. Keegan (1987:331) discounts one of the four sea routes Sears has postulated, but not the others.

A major argument against southern, or at least South American, origins has been a negative one, the absence of evidence for

[2]In supporting Sears's ideas regarding south-to-north migration, Riley says Sears "has reported some pollen evidence . . . in the Fort Center area of Florida at A.D. 200 . . . on linear ridges that Sears thinks are planting features." He cites a 1971 article in which I find no mention of such a date. Further, in his 1982 Fort Center book, Sears reports no radiocarbon dates for linear ridges and says they were in use only after A.D. 1200. The evidence for early maize comes instead from the Great Circle (Riley 1987:301; Sears 1971:322–329, and 1982:116).

75

cultivation of manioc, the major staple of lowland South America (Griffin 1943:89). D. D. Laxson has speculated (1966: 128) on the possibility of griddles at the Turner River Jungle Gardens site, but it is more likely that the sherds in question are merely bowl fragments. Widmer (1988:230) notes that no evidence of *Zamia*, another tropical root crop, has been found archaeologically in South Florida. The Fort Center excavations did nothing to change that conclusion (Sears 1982:119),[3] but what Sears did find is enough to tip the scales toward the south.

Sears also cites the fact that fiber-tempered pottery was made at Fort Center as evidence for southern origins. As noted in the previous chapter, Sears (1977:7, 9) rejects the conventional view that South Florida's fiber-tempered pottery diffused southward from the St. Johns River area; he says the Fort Center pottery "lacks decoration and the shape is wrong." Reports of fiber-tempered pottery from elsewhere in South Florida do not make a conclusive case either way. Randolph J. Widmer (1974:15) places the introduction earlier at Marco Island than on the east coast, and W. A. Cockrell (1970:60) reports Marco Island dates ranging from 1450 to 1140 B.C. As noted in the previous chapter, I am most comfortable with a 1500 B.C. date for the Palmer site in Sarasota County, though Ripley P. and Adelaide K. Bullen (1976:49) push it to 1700 B.C. All of these data are consistent with movement in either direction, as fiber-tempered pottery dates back to 3000 B.C. in Colombia and 2000 B.C. in the St. Johns area of northeast Florida (Ford 1969:12–13). The initial period at Fort Center falls squarely within the Transitional period. Jerald T. Milanich and Charles H. Fairbanks, speaking of Florida in general, sum up the rationale for the name *Transitional* nicely

[3]The archaeological evidence for manioc cultivation, according to Sears (personal communication), is in the form of the flat griddles used for cooking or the tiny flint chips that are driven into boards to produce graters. Some flint chips were found at Fort Center, but the closest thing to a griddle fragment aside from the Laxson find is some Glades Tooled rims that look a bit like they are from griddles (Milanich and Fairbanks 1980:234), and they date from well after the decline of the Fort Center ceremonial complex.

76

(1980:61) by calling it "a period of cultural transition from the hunting-gathering Archaic cultures to the many post–500 B.C. regional cultures."

To Widmer, speaking specifically of the Gulf Coast, the culmination of this process is related to the advent 2700 years ago of present-day sea levels and thus of present-day environmental conditions as "the sea level rose to a position which is optimal for the formation of highly productive coastal environments" (Widmer 1988:213). He cites "increased area of coastal ecosystems, increased productivity in these ecosystems as a result of increased sedimentation and water flow from the interior, and increased productivity of the freshwater aquatic swamp zone which now flanks the coastal zone." He goes on to say the changes were "so dramatic that we see a complete transformation in the cultural adaptation" (Widmer 1988:213).

For the first time, we can no longer speak of a single South Florida culture. By the end of the Transitional period the South Florida Archaic, as conventionally defined, has given way to a series of regional cultures. In recent years we have come to have a much better idea as to what those cultures were.

In the 1940s, John M. Goggin (1964b:79) defined as the Glades area all of Florida south of a line running northeast from Charlotte Harbor to intersect the Kissimmee River in the Basinger area, thence east to the Atlantic in the Fort Pierce area. To the north were the Manatee region on the Gulf Coast, the Kissimmee region in the interior, and the Melbourne region on the east coast. Goggin subdivided the Glades area into three subareas: "These are the Tekesta subarea, which includes the Florida Keys, the Cape Sable region, the lower Ten Thousand Islands, the Everglades, and the East Coast, south of Boca Raton. Northwards the Okeechobee subarea extends across the state to about 20 miles above Fort Myers on the Caloosahatchee River, and up the East Coast to just south of Stuart, or the Indian River. On the West Coast, the Calusa subarea includes the upper Ten Thousand Islands, the Big Cypress Swamp and the coast and islands up to, and including, Pine and La Costa Islands" (Goggin m.s.:69).

The most unfortunate aspect of this system was Goggin's use

of the names of historic tribes for prehistoric subareas. To compound the problem, the incised pottery types found in the Calusa and Tequesta (as it is now generally spelled) subareas have come to bear those names, also. As an example of the difficulties this nomenclature presents, consider that the types known as "Calusa"—notably Gordon's Pass and Sanibel—"rarely occur above Naples. . . . Ironically, it is in this northern portion that Mound Key, the best candidate for the historic site of the Calusa Village of Calos, is situated" (Ehrenhard et al. 1978:6–7). Further, as noted in a previous chapter, only one prehistoric site ever has been identified conclusively with the historic Calusa.

Goggin and William C. Sturtevant (1964:194–196) added to the confusion in the 1960s by saying that the Calusa had built such earthworks as the ones at Fort Center. In the first place, those earthworks were built well before there is any evidence of a culture even tentatively identifiable as Calusa. In the second place, even if the coast-dwellers of the period in question were Calusa, there is no reason to believe they were present in the Lake Okeechobee Basin.

Such confusions are not easily remedied. As recently as 1986, a local history book written by a university professor says, "The Calusa, a people from the lower southwestern coast, probably lived around Lake Okeechobee" (Curl 1986:9). While it seems indisputable that the historic Calusa held sway over the basin (Fontaneda 1973:30), the archaeological evidence suggests strongly that the lake people were not Calusa; they had a different subsistence base and used different pottery (McGoun 1981:23–33). Even more recently, a Broward County park dedicated in 1987 is billed as the site of Tequesta occupation dating back to 3100 to 2600 B.C. (Anonymous n.d.). As is the case with the Calusa, any attempts to link the historic Tequesta with ancestors going back that far are strictly conjectural.

There have been in recent years various reconsiderations of the Goggin scheme, all of which abandon the concept of a single Glades area. Robert S. Carr and John G. Beriault (1984:12) postulate a Lake Okeechobee area encompassing the lake basin as well as the Kissimmee River, bordered by wedge-shaped Caloosahatchee, Ten Thousand Islands, Everglades, and East

Okeechobee areas. The Ten Thousand Islands area is self-explanatory; the others comprise roughly Goggin's Calusa subarea, the Atlantic coast from Cape Sable north to Boca Raton, and the coast on north to the Martin-St. Lucie county line respectively. More recently, John W. Griffin (1988:121) has labeled as the Everglades area everything south of a line running eastward roughly from Bonita Springs on the Gulf to Palm Beach on the Atlantic, with a Ten Thousand Islands district within. His plan also includes a Caloosahatchee area and a Belle Glade area roughly the same as the Carr-Beriault Lake Okeechobee area. He leaves the Atlantic coast north of the Palm Beaches unclassified. The most adequate reconsideration is that of Milanich and Fairbanks. They (1980:22) have designated a Caloosahatchee area that includes the Ten Thousand Islands and have applied the name Okeechobee Basin to the Lake Okeechobee-Kissimmee River area. Elsewhere, they have lumped everything from Cape Sable north roughly to Melbourne into a single Circum-Glades area.

All of these reconsiderations have their advantages, including but not limited to removal of the historic-prehistoric confusion. Goggin himself said (1964b:88) the Kissimmee region "is a somewhat amorphous area tentatively considered," and subsequent work has shown no cultural distinctions between it and the Lake Okeechobee Basin. Most recently, a survey of the Avon Park bombing range has shown a preponderance of the crudely polished pottery known as Belle Glade (Austin 1987:291), which also predominates at basin sites (Goggin m.s.:458).

Where the Milanich-Fairbanks plan is superior is in its designation of a single area on the east coast; nothing in prehistory or history suggests any significant cultural differences within the area between Cape Sable and Cape Canaveral. Further, the total area considered by Milanich and Fairbanks defines, with the exception of the Gulf Coast from Charlotte Harbor north to Tampa Bay, the area in which gritty sand-tempered pottery predominates. Goggin, who defines this pottery as the Glades Series, notes (m.s.:416) that it "is found for some distance north of the Glades Area" on the west coast "and is usually in early horizons." The development of the west coast, especially in the

Caloosahatchee neighborhood but including the area north to Tampa Bay, will be considered in detail in the next chapter.

Goggin (m.s.:458) classifies the Belle Glade ware as a separate series, intermediate in terms of texture and paste between the Glades Series and the temperless St. Johns ware also found extensively in South Florida, but it can just as easily be classified as part of the Glades Series.

The one problem with the Milanich-Fairbanks system is the choice of the name Circum-Glades. Since their book came out, surveys of the Big Cypress National Preserve and Everglades National Park have revealed many "Circum-Glades" sites within the Everglades. For lack of a better name, I will simply re-christen the Milanich-Fairbanks Circum-Glades area as Southeast Florida. My scheme also distinguishes those areas that were subject to different influences at different times (see Figure 1).

Whatever it is called, by 500 B.C. the Okeechobee Basin was on a different path of development from the coastal areas. While the basin people were taking to horticulture, the others—with the possible exception of the people who built the circular ditches in Dade County—continued with the Archaic way of life. As Milanich and Fairbanks have noted, "The basic patterns of subsistence and settlement—villages at the mouths of rivers or on the coastal lagoon; seasonal occupation of smaller sites farther inland; hunting, fishing and gathering with no agriculture—persisted relatively unchanged into the Eighteenth Century" (Milanich and Fairbanks 1980:233). Only on the southwest coast could such a way of life support year-round settlements.

It is at 500 B.C. that the Transitional period is succeeded by the period known in the literature as Glades I. The advent of this period is defined by the presence of plain Glades Series sand-tempered pottery. This may be linked to the cultural changes going on in the Okeechobee Basin and on the southwest coast, which would strengthen the case for the importation of either people or ideas. Other areas accepted pottery and went on with life as before, at least as well as we can tell from the archaeological record. The sharp increase in the number of sites suggests a major population increase, though care must be used in assuming such a correlation.

Later in Glades I time there would be a lot more change. Early in the Christian era, the Fort Center farming community was converted into a ceremonial center, and "all the people on the site were directly involved" with the center's operation (Sears 1982:195). In other words, Fort Center had become the home of an elite group of people atop a society that was no longer egalitarian.

The fact that the bones discovered in the pond represent "a normal population distribution of ages and sexes" (Sears 1982:196) suggests that these people had high social status from birth; had status been acquired during life the bones would be expected to be generally those of adult males. Stephen Hale believes, basing his theory on differences in faunal remains, that the eating of deer meat well may have been a privilege reserved for the elite. However, his own data, which show deer as being plentiful all over the site, cast doubt upon that conclusion (Hale 1984:178, 183).

The complex consisted of a mound that was used both for ceremonies and as a residence, a smaller platform mound, and the pond with its wooden platform, all surrounded by a low earthen wall. As Sears reconstructs it (1982:145, 165, 168, 195), the platform mound was where "bodies . . . were probably roughly defleshed [and] made into bundles with a wrapping of matting" preparatory to interment on the wooden platform. The platform, like the pond, was in the shape of a "D" and apparently featured carved wooden birds and beasts placed atop its support poles. In some cases, the posts themselves were carved into figures.

All of this bespeaks presence of a mortuary cult. As Sears noted, this in turn implies the existence of religious beliefs. He believes further that maize cultivation may have been important in those beliefs and that the pond signifies that "water was a ceremonial requirement" (Sears 1982:187, 165). This last is consistent with the idea, discussed in previous chapters, that South Florida was the scene of a water mortuary cult extending back in time possibly to the Paleoindian period and forward until European contact. If the historic metal tablets in fact represent a caterpillar that attacks green corn, this could be another link between Fort Center at the time of Christ and Mound Key 1567 years later.

While the presence of burials in mounds shows that such a cult was not universal among people who had enough status to be accorded special funeral rites, it still might have been important over time in certain areas or for certain types of persons. In Glades I times it may have extended in space beyond Fort Center. Carr (1985:300) notes the presence of artificial ponds at two other sites in the Fisheating Creek area. Coleman (1989:262) sees a burial preparation area and a cemetery in a site near Tamiami Trail in Dade County, but the dating is uncertain.

Carr is not sure that the use of Fort Center changed as dramatically as Sears believes it did. "Sears' total rejection of any ceremonialism associated with the [Great] Circle is premature," Carr says (1985:299), because the sand mound near the center of the Great Circle could have had ceremonial significance.

Again as noted previously, Alice Gates Schwehm has studied the carved wooden objects from prehistoric South Florida. She sees (1983:112) a possible "soul in eye" concept from the big eyes found possibly at Fort Center and definitely at two other sites and thinks this belief may have persisted into historic times.

Milanich and Fairbanks (1980:184) note that birds were eaten only sparingly and speculate that this avoidance could indicate a taboo. This assumption would be consistent with Hale's observation (m.s.) that "the animals depicted in the wood carvings associated with the charnel platform tend not to be important as food items." It is, in fact, reasonable to assume that all animals thus depicted had ritual significance. Besides birds—the most numerous category—they include cats, foxes, and what look like dogs, though Sears believes they probably were bears (Sears 1982:42–55). Of course, there could be a more prosaic explanation for the paucity of birds in the diet. As Purdy (personal communication) notes, they are harder to catch than fish or turtles.

The Fort Center evidence points to a group of people who lived well, or at least ate well. Sears (1982:195) notes that the habitation mound yielded "tremendous quantities of bones of turtles, fish, deer, turkey, and other fauna." It is reasonable to

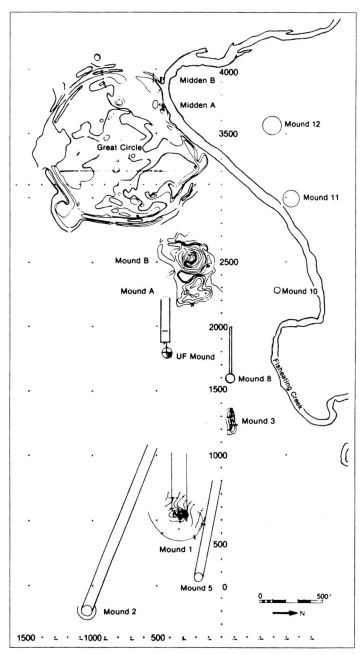

Figure 9. Fort Center (Sears 1982:4)

83

assume also that the people of Fort Center continued to exploit wild plant foods. In fact, the maize, while important ritually and as a reasonably stable source of food, probably was a relatively minor constituent in the people's diet. Carr believes (1974:23) that "hunting and gathering remained the substantial subsistence patterns for the area."

The best single-word explanation of what happened at Fort Center, and perhaps at other sites as well, is Hopewell. That word, however, conveys a range of meanings. Most anthropologists find it easier to describe the physical remains of Hopewellian sites—large earthworks and burial mounds, "objects of surpassing beauty and excellence" fashioned from exotic materials for adornment or ceremonial use (Spencer and Jennings 1965:64–67)—than to say what Hopewell is. Consider Joseph Caldwell's analysis: "It has been called a civilization, a culture, a complex, a phase, a regional expression of a phase, a period, a style, a cultural climax, migrations of a ruling class, a technological revolution, a social revolution, an in-place development out of previous antecedents" (Caldwell 1964:136).

Caldwell opts to call it "an interaction sphere embracing a number of distinct societies and separate cultures." He cites "striking regional differences in the secular, domestic and non-mortuary aspects . . . and an interesting, if short, list of exact similarities in funerary usages and mortuary artifacts over great distances" (Caldwell 1964:137–138). More recently, David S. Brose has argued that the mortuary items indicate status in an exchange network designed for the purpose of protecting individual societies against food shortages. Such a network would make more resources available to each society and would in turn both allow and encourage specializations and social stratification (Brose 1979:7–8).

Whatever it was, Hopewell persisted for six centuries or more centered on the time of Christ. While its strongest manifestations were in Illinois and Ohio, "Hopewellian influences were widest spread of any culture north of Mexico, save the Archaic, in all of North American prehistory," in the view of Robert F. Spencer and Jesse D. Jennings. They define (1965:64) that spread as east roughly to the fall line of the Atlantic coastal plain and west into Nebraska and Kansas, reaching the Gulf of Mexico between the Mississippi and the Suwannee rivers.

Edward McMichael believes the "Ohio Valley Hopewell was stimulated by Mexican influences and that the mechanism for this influence was the Crystal River Complex" of Florida's upper Gulf Coast. He postulates that the "complex of cultural traits" involved was introduced by sea from the Veracruz area of Mexico about the time of Christ (McMichael 1964:125, 131).

Sears does not seem prepared to accept the idea that Mesoamerican characteristics reached Fort Center by way of Crystal River rather than directly from the south, but he does see similarities between Crystal River and Fort Center: "It seems probable to me that the earth wall around the burial mound at the Crystal River site was similar in function to the one at Fort Center" (Sears 1982:198). Additionally, Sears (m.s.:15) accepts Caldwell's ideas in that he feels "Hopewell is not a culture but a tradition which is a part of many cultures."

Where Sears departs sharply from the conventional views of Hopewell is in rejecting Stuart Struever's idea that Hopewellian people subsisted on either intense gathering or cultivation of crops that today are not domesticated, especially certain grasses (Struever 1972:313). To Sears, the key word for understanding Hopewell is "maize."

Trade and apparent uniformity in ceremonialism, both over a wide area, are hallmarks of Hopewellian times. Spencer and Jennings report (1965:64, 66) that "Hopewellian imports to central Ohio include such things as alligator teeth and skulls from Florida, obsidian from Yellowstone Park or the Black Hills, mica from North Carolina or Arkansas, copper from the Lake Superior country, and conch shells from the Gulf of Mexico." Joseph Caldwell points out (1964:137): "A figurine from the Mandeville site in Georgia can be duplicated at the Knight Mound in Illinois. A cache of thousands of chipped stone 'blanks' occurs at the Baehr site in Illinois and again at the original Hopewell site in Ohio. Panpipes in Ohio, Illinois, and Georgia are virtually duplicated in Florida. Pottery vessels found as grave accompaniments in Illinois are nearly undistinguishable [*sic*] from vessels in Ohio and Louisiana."[4]

The wide variety of ceramics—"a hundred variations in vessel form and perhaps a thousand combinations of design and decorative technique over the Hopewellians' wide sphere of influence" (Spencer and Jennings 1965:66)—lend credence to the idea that Hopewell is a tradition grafted onto a number of otherwise dissimilar cultures.

Our understanding of this tradition is limited by the fact, as Spencer and Jennings put it (1965:66), that "the archaeological knowledge of the Hopewell is almost confined to the death cult and its furniture because few details of the day-to-day round have been captured." Milanich (personal communication) points out that excavations have been limited generally to burial mounds.

What Sears has in mind is a "system of corn agriculture with storage and preparation techniques which were tightly associated with religious and ceremonial concepts" (Sears m.s.:15). He goes on to stress the importance of those associated techniques: "[G]reen corn, corn on the cob, is not a food capable of supporting a culture. As the mainstay of an economic system it must be dried, stored, and then prepared as needed." He says that "Corn itself was probably around during most, or all, of the Late Archaic. But, without drying, storage, and preparation it had no real economic effect" (Sears m.s.:16–17).

As evidence that the Hopewell tradition was present at Fort Center, he cites the discovery of platform pipes, along with "a reasonably impressive total" of trade pottery sherds (Sears

---

[4]Spencer and Jennings say (1965:66): "[T]he most distinctive Hopewell complex is ceramic with the rocker stamp, the symmetry of design layout, the alternate use of smooth and textured zones in design, and the crosshatched rim as diagnostic features. . . . Some other diagnostic Hopewell artifacts . . . are the shell containers; alligator, shark and bear teeth; copper breastplates; drills and awls of copper; animal effigies on the monitor (platform) pipe and in stone and other materials; earspools of stone and copper; cut animal jaws; reel and bar-shaped gorgets; and cannel coal ornaments."

m.s.:35) and the lime. He also points to a "single ceremonial deposit, Hopewellian in nature" found at the edge of the platform mound nearest the pond (Sears 1982:187). Spencer and Jennings (1965:68) call the platform pipe "a striking example of the unique Hopewellian trait," while trade is acknowledged universally as a Hopewell hallmark. The lime is important in light of Sears's contention that a corn system is central to Hopewell. The paucity of maize finds at Midwestern Hopewell sites results, he believes (1982:199), from archaeological techniques rather than an absence of the cultigen.

There is some other evidence for maize early enough to have been a factor in Hopewell. Jefferson Chapman and Gary D. Crites (1987:352–353) report a date of roughly A.D. 200 from the Icehouse Bottom site in eastern Tennessee. Still the predominance of opinion is against Sears. Dee Anne Wymer (1987:211) believes the early maize dates reported for Ohio sites are due to historic or late prehistoric contamination.

The most commonly accepted view of Hopewell is that cultigens were grasses that are now considered to be weeds, a view stated most recently by Bruce D. Smith (1989:1569). In any case the presence of a cultigen does not necessarily indicate it was central to the diet. Domesticated plants could have been marginal both in the Midwest and at Fort Center. Wymer notes the possibility that maize was a ceremonial food in Ohio Hopewell (Spencer and Jennings 1965:66; Wymer 1987:211–212). This possibility is especially significant in light of Sears's view (m.s.:18) that the Green Corn Dance of the eastern historic Indians is a "survival of Hopewell ceremonialism."

As for the platform-mound cache, it is on the mound's east side and, as Sears has noted previously (1958:282), east-side caches are typical of Hopewell. The Fort Center deposit consisted of "a single adult human skull in poor shape, the skull cap of an infant with traces of cutting on its margins, several *Busycon* dippers, three *Venus* clam shells, a set of nested shells consisting alternatively of four clam and four scallop shells, two bird-bone tubes, and the cut and worked mandible of a small carnivore" (Sears 1982:157).

It is important to remember that these are the first mounds—

as opposed to middens, which are simply trash dumps—to have been constructed in South Florida. That in itself is an important link, if only stylistically, with the Adena and Hopewell populations of the Midwest, which traditionally have been identified in the literature as the Burial Mound people.

Sears believes (1982:197) that Fort Center may have functioned as a ceremonial center for a society that encircled the lake and extended up the Kissimmee River. While lending support to the extent of the society, the work of Carr and Hale in identifying other similar complexes suggests that Fort Center may have been only one of a number of roughly coequal sites. Unfortunately, we have virtually no data for any of the other sites, with one notable exception, the Belle Glade site, for which Sears himself sees the possibility of a role similar to that of Fort Center. It consists of a habitation mound and a burial mound that had been built in three stages. Human skeletal remains were found in all three stages (Sears 1982:197; Stirling 1935:374–376).

In addition to a wide range of pottery, the site yielded an abundance of stone, bone, shell, and wood artifacts. There were stone projectile points, weights or sinkers, tools, pipes, tubes, pendants or plummets; bone projectile points, daggers, pins, awls, adze sockets, punches, ornaments, headdresses; shell tools, dippers, ornaments; wood implements and effigies (Willey 1949:19–21). This last is especially interesting in light of the discoveries at Fort Center. Here again, birds were important; fragments of possibly five figures were found and the eyes appear to have been emphasized (Willey 1949:56–57). One bird from Belle Glade is identical to one from Key Marco (Purdy, personal communication). Another carving represents what Gordon R. Willey says is probably a dog but is more likely one of his alternate suggestions (deer, fox) due to the lack of evidence for dogs in aboriginal South Florida. Also, there are three plaques, one of which apparently contained a bird, and two human figures (Willey 1949:55–58).

Other human effigies have been reported for Lakeport near Fort Center (Willey 1949:78), and for Pahokee, on the east side of Lake Okeechobee (Purdy, personal communication). Willey

(1949:78) says the Lakeport object "resembles the two human figures from Belle Glade in style and proportions although better made."

Belle Glade is the type site for the crudely polished pottery mentioned previously. Based upon an analysis of that pottery, Willey (1949:71) identified two periods of occupation, though there is no break between the two. Stirling, basing his analysis on sequential use of first the habitation area and then the burial mound, identified (1935:375) three periods at each area. Milanich and Fairbanks believe (1980:186) the site was occupied from A.D. 500 into historic times.

That would make Belle Glade a bit late to have been subject to the ceremonial center at Fort Center. On the other hand, this connection provides a possible explanation of the human effigies, in that they may indicate a change in the religious system over time. Willey (1949:78) cites the Lakeport object's flattened skull, headdress style, and posture as "reminiscent of the larger stone figures, from other parts of the Southeast, done in the Middle Mississippian style." Anthropologists believe the Mississippian pattern—the name now used for the phenomenon known in Willey's day as Middle Mississippian—was in place by A.D. 900 (Chard 1975, 386), making Willey's observations consistent with the Milanich-Fairbanks dating.

Unfortunately, sites aside from Fort Center are either greatly disturbed or destroyed, which means there is virtually no chance of ever establishing conclusively what relationship there may have been with Fort Center. In fact, much of what we think we do know must be treated with caution. Many Glades I sites differ from pre-Glades sites only in the presence of plain sand-tempered pottery. In fact, the entire scheme of Glades periods, ranging from I (early) to IIIc (historic), is based solely upon variations in that pottery, almost entirely in the decorations of rims (Goggin m.s.:415). These differences may signify nothing more than a preference for a new style of rim.

Also, the seeming proliferation of sites in Glades I times— Widmer (1988:76) speaks of "a rapid expansion . . . in all areas of south Florida at this time"—may indicate nothing more than the relative ease with which occupation can be detected when pottery is present. After all, it appears that some South Florida groups accepted pottery because it was useful and went on with an opportunistic, egalitarian existence that was otherwise unchanged. Still, the changes detected for both the Okeechobee Basin and the southwest coast show that in at least these two cases the appearance of ceramics was part of greater changes.

The identification of some Glades I sites is suspect. Four decades ago, Goggin (m.s.:415) pointed out the importance of differentiating between rim and body sherds when reporting on Glades pottery, inasmuch as virtually all body sherds are plain, regardless of period. Unfortunately, this often is not done, which means that many "Glades I" sites may in fact belong to later periods.[5]

Still, when used with due care, the Glades Series does provide a useful means of dating many sites, determining sequences within and between sites, and detecting relationships with other areas. As noted before, it is the presence of Glades pottery that indicates the extension of Glades influence northward on the Gulf Coast to the Tampa Bay area—at least prior to A.D. 800—and on the east coast to the Cape Canaveral area.

Here again, Sears sees evidence of influence from the south. To him (m.s.:5-6) the Glades Series is part of a larger expression he calls the Florida Bowl Ceramic Tradition. "In this area . . . utilitarian pottery is characterized from its appearance to the historic period by simple open bowls. This . . . is really paralleled only in the Caribbean and in some parts of Lowland South America." Christopher Espenshade has a different view. To him the bowl shape is a function of necessity and not of cultural choice. Specifically, he says (1983:189-191) the material avail-

---

[5]One of the problems in determining the extent of settlement during Glades I times is illustrated in the fact that there are twenty-two South Florida sites listed as having been occupied in Glades I but not later. Unless there is a radiocarbon date, or some good stratigraphy, how can anyone be sure? The only other means of dating is through plain Glades ceramics, and in the absence of any overlying stratum containing decorated ware how is one to determine the age of the plain sherds?

able at the Gauthier site in Brevard County would have been poorly suited for more complex shapes due to its high shrinkage rate and lack of strength.

The Lake Okeechobee Basin was not the only scene of change during Glades I times. Widmer believes (1988:279) the advent of the Glades sequence also marks the beginning of a specialized adaptation to the rich marine resources of the Southwest Florida mangrove coast. This adaptation in turn set in motion the train of events that would culminate in the historic Calusa culture.

Somewhere between A.D. 600 and 800—or roughly at the start of the Glades II period—change came again to Fort Center. The ceremonial center was no longer in use and the people apparently were in transition to a new system of house mounds and linear raised fields (Sears 1982:199). Widmer believes (1988:278) a rising water table was making cultivation increasingly difficult. Another factor may have been what was going on in the southwest, a question that will be considered in detail in the next chapter.

FIVE

Down to the Sea and the Shells

The Shift of Power to Southwest Florida

About the time Charlemagne was unifying much of Europe under a single rule, a similar process was occurring in Southwest Florida. That process would give rise to the powerful Indian polity that killed Ponce de León and thwarted Pedro Menéndez de Avilés.

What happened, according to Randolph J. Widmer, is that the people utilizing the rich marine resources on the southwest coast had expanded in numbers to the limits of their environment, leading to the final step in a process of increasing social complexity that had been under way since the end of the Archaic. In this step, a dominant group of related people, known as a lineage, established political hegemony over the Southwest Florida coast and proceeded to annex inland areas in order to exploit their resources (Widmer 1988:279).[1]

Both the search for food and the search for water exerted various pressures that favored complex social organization, in Widmer's view. In a process involving positive feedback, control of productive fishing grounds by a kin group would increase their efficiency, which in turn would increase the power of the

[1]In precise terms, Widmer argues (1988:272) that Calusa social organization was a chiefdom of the unilineal-descent-group type, which had supplanted a ramage type about A.D. 800. The distinction, he says, is that daughter settlements retain their ties to the parent settlement in a ramage system whereas they do not in a unilineal-descent-group system. In the period of expanded settlement between A.D. 500 and 800, Widmer says, new communities frequently would not be self-sufficient and thus would need to retain ties to older communities. After A.D. 800, however, "when the area was filled in demographically and was environmentally circumscribed," this system would not be advantageous, "particularly since lineage members would probably be well represented in most of the villages scattered throughout the regions" (Widmer 1988:273).

kin group inasmuch as surplus production could be exchanged with neighboring groups. With this economic strength would come military strength as well (Widmer 1988:264). One benefit of territorial expansion would have been access to palm fiber for construction of fishing nets, again reinforcing the cycle (Widmer 1988:265).

Potable water was a scarce commodity, one that was husbanded through the construction of cisterns. Cushing (1973:336) speaks of "drainage-basins to catch rain for drinking water," and Hrdlicka (1922:25) notes a freshwater pond on a small island near Key Marco that could have been a cistern. Widmer says, "Construction of such features would require coordinated labor activity, and more important, would require the continual monitoring, maintenance, and cleaning of the cistern or pond to ensure the availability of a continuous water supply" (Widmer 1988:264).

The importance of fishing is demonstrated by analyses of various sites in the Charlotte Harbor area. Widmer notes that fish accounted for 97.7 percent of all biomass at Josslyn Island, 93.9 percent at the Wightman site on Sanibel Island, 92.6 percent at Caxambas on Marco Island, 92.1 percent at Useppa Island, 91.9 percent at Marco Midden, and 91.4 percent at the Solana site (Widmer 1988:236).

For the Wightman site, Arlene Fradkin says the most utilized fish were catfish, grouper, snook, jack, black drum, red drum, and hammerhead shark (Fradkin 1976:101). An intriguing anomaly is the general scarcity of bones from mullet, a plentiful food fish, a scarcity noted by John W. Griffin (1974:344) and Elizabeth Wing (1965:23) as well as Fradkin (1976:101). This finding could indicate that mullet were not common in late prehistory, that there was some cultural stricture against eating them, or that methods of preparation did not lend themselves to preservation.

Widmer believes that contemporary environmental conditions were established by roughly 700 B.C. He thinks that the coastal subsistence pattern could have been established within a generation and in any case is "clearly formed" by A.D. 280, citing as evidence the overwhelming predominance (94.7 per-

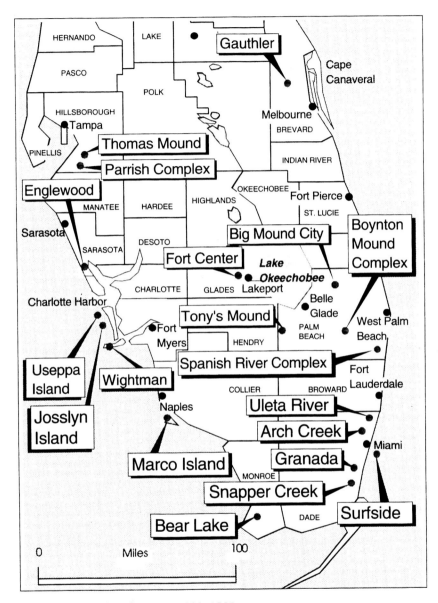

Figure 10. Major Sites, A.D. 800–1565

cent) of brackish and saltwater animals in Marco Island deposits. He says (1988:210, 215) the large size of two early Glades sites on Marco Island shows they were subject to "permanent year-round habitation by multiple-lineage groups."

To the north in Charlotte Harbor, Karen Jo Walker (1992:3) postulates that a canal could have been cut across Pine Island between A.D. 100 and 300, a time of high sea level during which "the canal would have been easier to dig." If she is right, this is evidence of social organization strong enough to manage a major public-works project.

Rapid colonization of the Southwest Florida coast soon followed, Widmer says, until all good sites were occupied. This, he says, occurred by A.D. 800, due to the fact that "Almost *all* sites [in the area] have components dating from A.D. 800 on, while only some have earlier components" (Widmer 1988:215–217). At this point ensued the process of hegemony and expansion cited above.

The importance of contact with the interior is shown by the prehistoric canals across Cape Coral and at Ortona on the Caloosahatchee River (Luer 1989:126). What the coastal people had to offer in exchange is clear; it was the surplus from their enormously productive estuarine areas. What the interior had to offer is less clear; William H. Sears (1982:193, 195, 199–200) believes maize was cultivated throughout the occupation of the Fort Center site, whereas Stephen Hale (m.s.) thinks maize gave way to root crops about A.D. 500. For his part, Widmer (1988:278) believes rising water levels made maize cultivation impossible in the Lake Okeechobee Basin by A.D. 1000. Grinding stones have been reported at Marco Island and at Lakeport (Small 1929:76; Carr 1975:16), but they do not necessarily indicate cultivation, as they could have been used to grind wild foods.

Still, there must have been some cultigens available, along with wild plants and inland animals. Of these, however, only the roots were of much importance in terms of exchange with the coast, in Widmer's view (1988:274). He says (1988:275) the coastal groups would have offered shell and shark teeth, which is consistent with Wing's view (1977:54) that marine resources were used for food only on the coast and that only nonfood items were traded.

The matter of coast-inland relationships remains a field for conjecture. For instance, could it be that the complex societies of the southwest coast were founded by migrants from the Lake Okeechobee Basin? Alice Gates Schwehm (1983:110, 112) theorizes about a lake-to-coast diffusion of both art styles and people. Fradkin notes (1976:103) the pottery similarities, in that both the basin and the Caloosahatchee area are marked by undecorated ware. Milanich and Fairbanks speculate (1980:181) that "Later development of the Caloosahatchee region may have been due to the . . . peoples occupying the basin." Perhaps, but the archaeological record for the Gulf Coast does not show the sort of discontinuity that would be expected as the result of an intrusion significant enough to effect a major change.

Another subject for conjecture is the relationships between the mangrove coast from Charlotte Harbor south through the Marco area and the coast northward to Tampa Bay. The period during which the southern area was developing, according to the Widmer chronology, has been designated as the Manasota period for the coast northward. George Luer and Marion Almy postulate (1982:52) a coastal adaptation less intense than to the south, allowing "people to reside on the shore for much of the year and to stay for short intervals at sites inland."

The artifacts of this early period suggest at least some interaction with areas to the south and east. Goggin noted (m.s.:416) that gritty sand-tempered pottery is found on the west coast "for some distance north of the Glades Area and is usually in early horizons." He speculated on a diffusion north-to-south. Diffusion in the other direction is suggested by the presence at a number of sites—Thomas Mound, Cockroach Key, and Picknick in Hillsborough County; the Parrish complex in Manatee; and Englewood in Sarasota County—of significant quantities of the crudely polished gritty ware known as Belle Glade (Willey 1982:120–121, 133–134, 144, 150, 156, 165, 336).[2]

---

[2]In addition to the problem of identifying a prehistoric pottery type with a historic culture, there is the problem of using a site name that may be inappropri-

Wares of the Safety Harbor series, named for a Tampa Bay site that is believed to have been the capital of the historic Tocobaga people, are widespread in South Florida but generally in small numbers, which indicate trade rather than the presence of outsiders. Widmer (1988:279) sees the extent of Safety Harbor ceramics in South Florida as evidence of Calusa hegemony. Jeffrey Mitchem is not so sure. He says (1989:577) that more archaeological work is needed on inland sites south of Charlotte Harbor "to determine whether Safety Harbor . . . sherds reported from these areas represent actual Safety Harbor occupations or exchange with Safety Harbor groups."

The existence of a nonegalitarian society on the southwest coast by A.D. 800 seems beyond question. In addition to Widmer's observations about site distribution there is the list of burial mounds he recounts (1988:162–165) ranging from Charlotte Harbor north into Manatee County—more evidence suggesting ties throughout the area in question. And then there is the mound complex reported by Fradkin (1976:102) for the Wightman site, a complex that Widmer feels clinches the case. Milanich (personal communication) thinks the Collier mound on Useppa Island offers further evidence. Fradkin (1976:103) defines the key area as stretching from Mound Key north to Charlotte Harbor, which means it corresponds closely to the historic Calusa heartland and thus strengthens the idea that the people living there were either the Calusa or a group ancestral to them.

Elsewhere in South Florida, the picture is less clear. The people at Fort Center seem to have been in another transitional

---

ate. For instance, Sanibel Incised is rarely found on Sanibel Island, and the Fort Drum for which another type is named is not the reasonably well known Fort Drum community in Okeechobee County but rather the virtually forgotten Fort Simon Drum, a Second Seminole War encampment deep in the Everglades. In the case of Belle Glade pottery, the record is ambiguous. It was predominant at the Belle Glade site throughout that site's occupation, but it also predominates at Fort Center after A.D. 500, the apparent beginning date for Belle Glade occupation (Willey 1949:62–68; Sears 1982:112).

phase, one that Sears says (1982:199) is known primarily by what is not known. The lack of evidence at Fort Center gives support to the possibility that dominance within the Lake Okeechobee Basin had passed to the Belle Glade people.

The strongest evidence for social complexity on the east coast is the Barnhill Mound site at Boca Raton. Ripley Bullen (1957:34–36) believed the mound was built in three stages, with the first predating A.D. 700 and the third at roughly A.D. 1300. John Furey (1972:101) sees evidence of a ranked society in the careful preparation of the first stage. Furey also argues (1972:95) that agriculture was practiced in this area. In the absence of evidence, however, his theory, which rests on analogy with Fort Center and with Boca Raton in the twentieth century, must be considered tentative.

In fact, just about any conclusion regarding the east coast must be carefully qualified. The area has been subjected to the most intense twentieth-century development, and therefore the greatest destruction of archaeological sites, of any area in South Florida. There may have been many areas with more to offer than the Barnhill Mound and several nearby sites, all known collectively as the Spanish River complex (Furey 1972:89), but they were destroyed before they could be examined in any detail.

One good argument in favor of the superiority of the mangrove-coast subsistence base, and by extension against the presence of agriculture virtually anywhere else, is the fact that almost all occupation elsewhere seems to have been seasonal. Fradkin (n.d.:15) has noted a predominance of winter birds at Boca Weir, part of the Spanish River complex, which could indicate the site was not occupied in the summer.

This idea would be consistent with zooarchaeological analyses at the Granada site in Dade County. C. Margaret Scarry says (1982:232–233) that while a "variety of plant foods would have been available . . . year round" that "the same set of resources was found in all the samples. For this to occur so consistently, the various plant foods would have to have been collected over a relatively short time span within the year." An analysis of the availability of twenty-three vegetable foods, mostly berries and

99

fruit, shows that the time span in question would have been in the fall. From these and other data, John W. Griffin concludes (m.s.:386, 388) that the Granada site was occupied during the fall and winter, with the inhabitants utilizing interior resources in the summer.

Ponce de Leon visited Biscayne Bay twice in 1513, on May 13 and about July 3. The account by Antonio de Herrera makes no mention of Indians, suggesting that the village called Cheque-scha in his account, probably the Granada site, was unoccupied (Davis 1935:19, 21).

The typical subsistence pattern is described succinctly by Lewis H. Larson, Jr. He says animal resources in general and marine resources in particular were vital because "none of the important plant species that were collected in southern Florida were a source of vegetable oil. . . . The coastal groups seem to have focused their subsistence efforts on marine animals: shell-fish, fish, and sea mammals. The gathering activities of these groups were limited by the seasonality of the few species that were available for exploitation" (Larson 1980:223, 227). Larson believes at least some gathering went on year-round but that "a period of concentrated plant collecting took place in the fall when the saw palmetto berries, yucca, cocoa plums, sea grapes, and prickly pear ripened" (Larson 1980:223).

Griffin has argued (m.s.:8–9) that the distinction between shell and earth middens is seasonal: "It is postulated that the large shell sites represent the major base camps or villages, occupied during an undetermined portion of the year which almost certainly included winter. Summer occupancy is evidenced in some of the larger black earth middens by the heavy concentration of sea turtle bones. The small middens in the Everglades represent short-term hunting expeditions into the Glades, possibly in the spring-summer wet season when the deer concentrate on and near the tree islands."

The period beginning at roughly A.D. 800, known to archaeologists as Glades II, seems to have been of increasing interaction throughout South Florida. Griffin (m.s.:19) has noted that artifacts, especially those made of bone and shell, suggest an increase over time in trade between the Lake Okeechobee Basin

and areas to the south. Belle Glade pottery shows up increasingly in such coastal sites as Boca Weir (Furey 1972:39). The
differences lessen between east coast and west coast pottery
incision styles—between what Goggin (1964:86) has called
simple repeated designs on the east coast and more complicated
designs on the west (see Figure 11)—indicating increased contacts. John Ehrenhard and his co-authors (1978:7) believe these
contacts were in the form of intermarriage with the woman,
who he assumes was the potter, going to live with the man's
group. Maybe, but it would be just as plausible to postulate the
movement of the idea rather than the potter.

One direction in which contacts seem to have lessened, however, is between the Glades area as conventionally defined and
the Manasota area. The Manasota area came under the influence
of first the Weeden Island culture and then the Safety Harbor
culture, both emanating from farther north (Luer and Almy
1982:52).

By A.D. 1100 the expansionistic people of the Charlotte Harbor-Estero Bay area seem to have extended their sway all the way
to the southern tip of the state. Griffin (1988:140) points to the
virtual disappearance of decorated pottery styles in favor of the
undecorated pottery that was throughout prehistory the dominant style of the historic Calusa heartland.

About the same time, there was another change at Fort Center. Sears notes (1982:200) the introduction of a large amount of
Belle Glade pottery along with what he sees as the development
of raised-ridge farm plots tended by families living atop small
house mounds. If he is correct, the Fort Center development
may be just one manifestation of a late prehistoric farmstead
culture centered on Belle Glade. Similar structures are reported
for Tony's Mound (Allen 1948:17), Big Mound City (Willey
1949:73), and the Boynton Mound complex (McGoun 1987). All
five sites, along with others mapped by Hale (1984: Figure 1),
could have been in fairly easy contact with each other. Fort
Center is on Fisheating Creek, the Belle Glade site was on the
since-obliterated Democrat River, which emptied out into the
Everglades southeast of Lake Okeechobee (Will 1977:44–45), and
the other three are on the margins of the original Everglades.

101

Figure 11. Some Incised Pottery Styles—*top left:* Gordon's Pass; *top right:* Sanibel; *middle left:* Fort Drum; *middle right:* Surfside; *bottom left:* Glades Tooled; *bottom right:* St. Johns Check Stamped (Mowers 1975:38–39, 41)

The sites may have represented a centralized polity, but it is more likely they represented merely a common culture given the apparently small population of each. (Sears believes [1982:200] only one or two families at a time lived at Fort Center.) Possibly the sites other than Belle Glade represent daughter settlements created through fissioning; the presence of a burial mound at Belle Glade suggests it could have been a ceremonial center for those other sites. A further possibility is that the Boynton complex could have been in contact with Boca Weir, which is only some ten to twelve miles distant. Perhaps the Boca Weir people occupied the Boynton complex only in the summer.

Ryan J. Wheeler (1992:12) hypothesizes links inland from sites in northern coastal Palm Beach County that he calls the Riviera complex. He sees parallels to Sears's raised ridges, but an illustration with his article (Figure 1) indicates the structures were much more irregular than those reported elsewhere. Also, the "causeway" described by Goggin (m.s.:383) seems too high to have been built for agricultural purposes.

To call these sites farmsteads is not to imply that maize necessarily was the staple food. The raised ridges were too small to have yielded enough food for that purpose; more likely maize was cultivated as a supplement to hunting, fishing, and gathering. Also, no maize pollen has been found on any of them. Alternatively, maize may have been a ceremonial food. The largest single find of maize pollen at Fort Center was in the white pigment associated with a wood carving (Sears 1982:122), suggesting ceremonial significance there. As noted in the preceding chapter, Sears believes the historic Green Corn Dance may have had its origin in Hopewell ceremonialism 2000 years ago.

In any case, the cultivation could not have lasted into historic times, or else there would have been some mention in Spanish accounts. The most likely explanation is that the ridges were an adaptation to increasingly unsatisfactory environmental conditions, specifically the rising water levels of which Widmer speaks, that continued to become more inhospitable until maize was abandoned entirely. The small size of the sites suggests this lifestyle was not feasible for large groups. Another possibility is that the farmsteaders represented yet another migration from the south, bringing a subsistence system that was doomed from the start because it was ill-suited to Florida conditions.

While it is hazardous to reach too far in equating ceramic styles with cultures when those styles are simple, the parameters of the Belle Glade style are interesting. The distribution is centered on the type site and at Fort Center, the only sites with good time controls. At Fort Center, it becomes abundant late enough in time (Sears 1982:112) to allow for diffusion from Belle Glade.

Goggin notes (m.s.:458) that "The hearth of this ware is apparently in the Okeechobee subarea" and that it is found at sites ranging from Charlotte Harbor to the Cape Canaveral area. Put that with the previously noted reports for the Gulf Coast north of Charlotte Harbor, the Kissimmee River Valley, and the east coast at Boca Raton, and you have a distribution that includes much of South Florida.

Unfortunately, ceramic evidence from the sites on the rim of the Everglades is sparse. There are no reports for either Big Circle Mound or the Boynton Mound complex. Belle Glade ware does predominate at Big Mound City, but the sample size is small (30 sherds) (Willey 1949:76).

About A.D. 1200, at the beginning of what is known as the Glades III period, there appear to have been some significant cultural changes, the nature of which remains obscure, throughout South Florida but especially in the Southeast Florida area. At the moment, the only evidence is ceramic distributions.

Most notable is the presence in large amounts of the chalky, temperless ware known as St. Johns because of its dominance in Northeast Florida (Goggin 1964b:92). Most writers have assumed the chalky ware found in South Florida was brought in from the north—as recently as 1982, Sears (25) classified it as trade ware—but Espenshade has made a convincing argument that it was locally produced. Espenshade argues that "the pan-state distribution of chalky sherds is related to local production using locally available mucks." In support of that contention he cites evidence from both Fort Center and the Gauthier site in Brevard County and "a correspondence between areas yielding chalky sherds and the distribution of mucks" (Espenshade 1983:185–186).

Michael Russo feels much the same way. Further, he says (1985:2) this "temperless" ware in fact is tempered by sponge spicules, which produce the "chalky" feel as they are sloughed off. The evidence, he says, "points to local manufacture of ceramics from widely dispersed clay sources containing sponge spicules." Espenshade (1983:186) sees the suggestion of "a general South Florida pattern . . . indicative of the use of local mucks in their untempered and sand-tempered states." Taken to its extreme, such an argument could be used to claim that temperless ware evolved for environmental reasons, specifically the scarcity in late prehistory of muck deposits intermixed with sand.

That, however, would be stretching it a bit too far. Diffusion, at least of ideas if not of wares, is suggested strongly by the appearance in South Florida during Glades III times of large quantities of check-stamped temperless ware, which became dominant in the St. Johns area about 200 years previously. Additionally, incised pottery virtually dies out, with the exception of what Goggin (1964b:87) considers "a very simple form, Surfside Incised," on the lower east coast.

The most intriguing aspects of Glades III, however, is the decrease in the number of sites. In the Big Cypress National Preserve the decrease is from 33 to 20. For Collier County as a whole the decline is even sharper, from 43 sites to 25 (site cards, Bureau of Historic Preservation, Florida Department of State; site data, Southeast Archaeological Center, National Park Service, Tallahassee).

"All the larger midden sites on Pine Island [the Broward County hammock, not the Charlotte Harbor island] date from the Glades II Period," and no Glades III sites have yet been found (Carr 1990:260). In Dade County, several seemingly choice coastal locations (Snapper Creek, Arch Creek, Surfside, and Uleta River) were abandoned, and there is a general paucity of check-stamped chalky ware south of Miami River (McGoun 1975:4–6). Various suggestions have been put forward about the former sea-level fluctuations, salinity changes, cataclysmic storms, or, in the case of Arch Creek, the depletion of clay deposits for making ceramics (McGoun 1975:4).

Could rising sea levels account for the abandonment? Probably not. Taylor R. Alexander (1974:219–221) argues on the basis of research in the Keys that sea level has been rising in recent times, but he speaks only of the last century or so. Rhodes W. Fairbridge (1974:227) charts sea-level fluctuations that show a rising trend from A.D. 1000 to 1400, but the total change in that time was less than three feet.

The crowning glory of pre-contact South Florida comes, interestingly, at one of the first sites to be discovered. This is the Cushing site on Marco Island, commonly called the Court of the Pile Dwellers. In 1896, Frank Hamilton Cushing found an extraordinary array of wooden objects preserved in the site's muck (Cushing 1973:358), as well as objects of shell, bone, antler, stone, and cordage.

Marion Spjut Gilliland has cataloged these items. The wooden assemblage is especially important in that it points to cultural continuity both forward and backward in time. The items include a number of animal figures with the eyes emphasized (Gilliland 1975:Plates 61–64, 67–69, 73), suggesting links to the Okeechobee Basin tradition of an earlier period. And there is a wooden example of the sort of metal tablets found all over the Calusa sphere of influence in historic times (Gilliland 1975:Plate 35); this shows that the symbolism of the objects clearly was aboriginal in nature and further that the Key Marco site almost certainly was under Calusa domination. A carved cat on a bone pin, bearing some similarities to the one executed in wood at Key Marco, was found in south Dade County (Lord 1989:263–264).

Widmer believes, on the basis of stratigraphic data from Marco Midden, that this is one more of the major coastal sites that were home to nonegalitarian settlements by A.D. 800. That would seem to conflict with Milanich's assertion (1978a:682) that it is a single-component site with a "temporal range of late 15th and 16th Centuries." Milanich (personal communication) sees no conflict; he says the shell mound undoubtedly is as old as Widmer believes but that most of the Cushing site artifacts are protohistoric. Griffin (1989:197) says, "We do not know the time depth of the Key Marco wooden types and styles. But we do know that the site has very considerable time depth."

There is other dissent as well. Purdy (personal communication) believes the component is older, and her point that one bird head from Belle Glade is identical to one from Key Marco supports that view. Griffin (1988:1315) also argues for greater antiquity at Key Marco and says therefore the material found there cannot be assumed automatically to have been Calusa.

I believe it is Calusa. The wooden tablet is so like those made of historic-period metal that I cannot consider it to be of great antiquity. The similarity of Key Marco objects to those found elsewhere in South Florida would indicate the persistence of those other sites rather than an earlier occupation for Key Marco.

One point is unequivocal. The cacique who confronted the conquistador that February day in 1566 had a lot of tradition behind him. He probably represented a way a life, if not a polity, that dated back at least seven centuries.

SIX

The Road to Extinction

Aboriginal Peoples after the Menéndez Period

After the 1560s, neither the Spanish nor the Indians saw anything to be gained by further Spanish settlement in South Florida. Each could get what it wanted from the other without the necessity of living together. As a result, the historical record becomes sparse. The discovery of European artifacts at sites such as Nebot in Palm Beach County (Kennedy and Iscan 1987) indicates merely that the people there had contact with someone who had obtained such objects. The main resource of South Florida was its fish, which the Spanish could harvest by setting up seasonal camps on offshore islands roughly from August to the following March. James W. Covington (1959:117) says thirty or more Cuban vessels were operating in South Florida in 1770 and speculates that the custom dated back to at least 1600.

Indians could obtain European goods from shipwrecks, from visits to these offshore "ranchos," or by the sea journeying to Cuba. "Many of these Indians traveled from the mainland to Havana in their small canoes, making the journey from the Keys in twenty-four hours, and traded with the people in the city and the sailors. They carried fish, ambergris, tree bark, fruit, and a few hides or furs. . . . A very profitable item was the sale of cardinal birds to the sailors." The Indians seem to have taken away mostly cash to be used elsewhere, rather than Cuban goods (Covington 1959:116).

Writing of his experiences after a 1722 shipwreck, Pierre de Charlevoix tells of an Indian village on the Keys whose inhabitants traveled to Havana annually and supplemented their fishing with salvage (Charlevoix 1966:326, 329). On the Tortugas the Charlevoix party found what seems to have been a rancho abandoned for the summer (Charlevoix 1966:334). Post-Menéndez presence at Rookery Bay in Collier County is suggested by the discovery of two fragments of Chinese porcelain

(Goggin 1950:230). Marquardt (1987b:9) says contacts between
Florida and the Bahamas may have been established in the last
century before European contact and probably increased there-
after.

All in all, the European influence on the basic aboriginal
cultures in South Florida seems to have been minimal, aside
from the demographic effects of disease. And even here, later
accounts suggest the depopulation was not as great as has been
reported for other areas. Covington (1959:116) says the Indians
who visited Havana were little affected by Cuban culture, and
Elizabeth S. Wing (1978:7) says animal remains excavated at the
McLarty site in Indian River County do not reflect any differences in subsistence before and after European contact. Goggin
(m.s.:56) reports "a suggestion that the Calusa developed a form
of pictographic or other writing in the 17th and 18th centuries,"
but the matter apparently has not been explored further.

The Calusa seem to have maintained their power for a century
and a half after Menéndez. In fact, they may have expanded
their influence in the decades immediately afterward. Swanton
(1922:343) quotes Woodbury Lowery as saying they controlled
70 towns in 1612, and Mooney says they numbered 3,000 as late
as 1650 (Fairbanks 1957:44). Marquardt (1987a:108, 1988b:185)
sees a reconsolidation of the Calusa following the decentralizing
influences of Spanish wealth during the Menéndez era. Michael
Gannon (1965:69) speaks of continued hostility in 1678,
Swanton (1922:343) says the Calusa still were strong in 1680, and
Andrés Gonzáles de Barcia Carballido y Zúñiga (1951:344) says
they drove off friars in 1697.

Newly discovered documents give more detail about this last-
mentioned episode. "According to one document, there were
about 14,000 Indians in the Calusa province in 1692, some 2,000
of whom lived on the island where the chief's village was located.
The chief's village itself had 16 houses" (Marquardt 1988a:4).
The story of these clergymen and their short-lived attempt to
establish a new mission among the Calusa has a familiar theme.
The clergymen, here identified as four Franciscan priests, ran
afoul of the Indians when they first refused to distribute food and
clothing and then tried to supplant the Indian religion. They

110

were given a canoe and two small boats and sent on their way, finally being found naked and starving after being robbed by other Indian groups along the way toward Havana (Marquardt 1988a:4).

The 1697 experiences sound a lot like those of Rogel 130 years earlier. The cacique told one priest that "if he was not going to give clothing and food, it was not good to become a Christian and no one would want to become one." The priests were beaten when they persisted in going to the Indian temple (Friars, as translated in Hann 1991:166–167). In speaking of the temple, Feliciano Lopez describes "a sort of room (*aposento*) (made) of mats (*esteras*)" upon a "very high flat-topped mound. . . . The walls are entirely covered with masks, one worse than the other" (Feliciano Lopez, as translated in Hann 1991:159–160).

Similar cultural persistence is evident among the Ais and Jeaga, on the basis of Dickinson's 1696 experiences. As for their relations with the Spanish, there is a difference of opinion here. Lyon (1967:11) says the Ais remained continually hostile, and Hale G. Smith and Mark Gottlob (1978:8) tell of a Jeaga uprising in 1618, while Irving Rouse (1951:55) says relations were generally amicable after 1605. Perhaps Lyon and Rouse are looking at intermittent hostilities in two different ways. Still, there were indications of disruption. Calderón says that as of 1675 no South Florida group, not even the Calusa, had year-round villages (Wenhold 1936:11), and Swanton (1922:343) says Guale Indians from Georgia had settled among the Calusa in 1681.

After about 1700 the situation changed quickly with the advent of raiding southward from English territory, by both the English themselves and by Indians. Feliciano Lopez says the son of the Calusa chief in 1697, who apparently held the real power, spoke "half in Timucua and half in Apalachee," both languages of North Florida (Feliciano Lopez, as translated in Hann 1991:160). In this light, it is understandable why the Indians met by the Dickinson party disliked the English so much. Such Anglophobia also was noted in the accounts of shipwreck victim Briton Hammon (1760) and of Charlevoix (1966).

The new Indians, some of them ancestors of today's Seminoles, had reached as far as Ais territory by 1703 and the Keys by

1708, attacking mainland aborigines who had taken refuge there and forcing the Spanish to evacuate many of them (Rouse 1951:58; Wright 1981:142, 115). In 1711 a Spanish captain brought as many as 270 Indians to Havana, "and he stated that he would have brought more than two thousand had he had the vessels" (Valdes, as translated in Hann 1991:337). Charlevoix (1966:335) says few Indians were on the Gulf Coast by 1722. A lesser number of Indians evidently fled north toward St. Augustine and San Marcos de Apalachee (Hann 1991:357).

The fact that many Indians picked the Keys for a refuge indicates how hard they were pressed. Consider Charlevoix's description (1966, 329–330) of the land ("There is not so much as a single four-footed beast on these islands, which seem to have been cursed by God and man") and of a chief ("He is very near as naked as his subjects. . . . [T]he few rags on his back were hardly worth picking up at one's feet. . . . He was without attendance, or any mark of distinction or dignity"). Granted, Charlevoix probably was influenced by his plight, but the Keys were a hard place to live.

Whether the decline was caused by the raiders directly, through either enslaving or killing the Indians, or indirectly through disease, it was rapid. Rouse (1951:58) says he can find only one mention of the Ais after the early years of the eighteenth century. That mention is within the document that is virtually the only source of date about mainland South Florida in the mid-eighteenth century: the report of Fathers Joseph Maria Monaco and Joseph Xavier Alaña, who in 1742 made at present-day Miami the only known post-Menéndez attempt at establishing a Southeast Florida mission. The priests tell of a village of no more than five huts "in which up to one hundred and eighty people were living crowded together between men, women and children, the [latter of] which made up about half this number. . . . And they are all the remnant of three nations, Keys, Carlos, and Bocarraton. We learned that from another three tribes in addition to these, the Maymies, the Santaluzos, and the Mayacas, which have united [and are] four days' journey away on the mainland, it will be possible to add another hundred souls or a few more" (Alaña, as translated in Hann

1991:420). In other words, roughly 300 Indians native to South Florida still remained there.

These Indians would not permit their sons to be punished but would make sacrifices of children upon the death of the chief or "other leading men" and as part of celebrating peace (Alaña, as translated in Hann 1991:423). Their burial rites were not reflective of a belief in an enduring soul (Alaña, as translated in Hann 1991:424), a notable difference from Calusa customs of the Menéndez era as reported by Rogel, and "They have a great fear of the dead" (Alaña, as translated in Hann 1991:423). The wooden "idol" that the priests called "the God of the Cemetery" was carved with the image of a bird (Alaña, as translated in Hann 1991:422), perhaps reflecting continuity far back into prehistory, a point discussed in earlier chapters. Alaña considered the bird hideous and had it smashed and burned, along with the hut in which it was displayed. Two centuries previously such an outrage probably would have led to a full-scale attack on the Spaniards, but these embattled remnants evidently were not prepared for any action aside from "many signs of grief and even lamentations and tears from their women" (Alaña, as translated in Hann 1991:422).

Alaña also speaks of "the most ugly mask destined for the festivals of the principal idol," suggesting that at least some things had not changed since Rogel's time (Alaña, as translated in Hann 1991:422).

The plight of the Indians is described graphically: "These diminutive nations fight among themselves at every opportunity and they are shrinking as is indicated by the memory of the much greater number that there were just twenty years ago, so that, if they continue on in their barbarous style, they will have disappeared within a few years either because of the skirmishes, or because of the rum that they drink until they burst, or because of the children whom they kill, or because of those whom the smallpox carries off in the absence of remedies, or because of those who perish at the hands of the Uchises," one of the intruding groups from the north (Alaña, as translated in Hann 1991:427).

The priests set up a temporary outpost and recommended that

113

a permanent presence be instituted, complete with a military garrison and settlers. Their pleas fell on deaf ears, however, and the outpost soon was abandoned and destroyed (Güemes y Horcasitas, as translated in Hann 1991:409).

When Great Britain took possession of Florida in 1763, marking the end of the first Spanish period, the last of the Calusa went to Havana with the departing Spanish, according to Bernard Romans (1940:291). (Sturtevant [1962:70; 1978:141] places the number of evacuees at eighty families.) Some argue that Romans was wrong, that Calusa remained in South Florida and even maintained a separate identity. Some could have fled north, but that is unlikely.[1]

Several writers have said the village of Muspa existed until 1800 (Brinton 1859:114; Hodge 1911:963; Williams 1962:33), and two have said that it persisted through the time of the Second Seminole War of 1835–43 (Safford 1919:431; Douglass 1885:281). However, both Sturtevant (1953) and Wilfred T. Neill (1955) have considered the question of the "Spanish Indians" who fought in the war and concluded that they were migrants from the north and not Calusa holdovers. Neill quotes heavily from historic sources in indicating that the Spanish Indians knew that language because of either working with Spaniards at the fishing *ranchos* or operating the *ranchos* themselves and trading with the Spanish, but that they clearly were Seminoles and not Calusa.

It is hard to see how anyone could be certain no individual South Florida aborigines remained after 1763, given the inaccessibility of the interior. Still, even if there were a few Calusa among the Seminoles, the point is that 10,000 years of aboriginal tradition had ended, as T. S. Eliot once said in a different context, "not with a bang but a whimper."

---

[1]The report of a number of South Florida objects from a site in St. Marks in Wakulla County (Goggin 1947:273–276), south of Tallahassee, could be evidence of the Calusa visit to the Apalachee area of North Florida in 1688 (Hann 1991:78). However, the objects in question were in the private collection of an avocational archaeologist, and it is not certain they were curated carefully, raising the possibility that they actually were found in South Florida.

114

Bibliography

Alaña, Joseph Xavier
1991   "Report on the Indians of Southern Florida and Its Keys by
            Joseph María Monaco and Joseph Javier Alaña Presented to
            Governor Juan Francisco de Güemes y Horcasitas, 1760." In
            Missions to the Calusa, edited and translated by John H.
            Hann. Gainesville: University of Florida Press.

Alexander, Taylor R.
1974   "Evidence of Recent Sea-Level Rise Derived from Ecological
            Studies on Key Largo, Florida." In *Environments of South
            Florida: Present and Past,* edited by Patrick J. Gleason.
            Miami Geological Society Memoir 2.

Allen, Ross
1948   "The Big Circle Mounds." *Florida Anthropologist* 1:17–21.

Allerton, David, George M. Luer, and Robert S. Carr
1984   "Ceremonial Tablets and Related Objects from Florida."
            Florida Anthropologist 37:5–54.

Andrews, Charles M.
1943   "The Florida Indians in the Seventeenth Century." *Tequesta*
            3:36–48.

Andrews, Rhonda L., and J. M. Adovasio
1988   Textile and Related Perishable Remains from the Windover
            Site. Paper presented at the 53rd annual meeting, Society
            for American Archaeology, Phoenix, Arizona.

Anonymous
n.d.   *Peace Mound Park, Indian Plant Identification, Trail Guide.*
            Pamphlet.

Austin, Robert J.
1987   "Prehistoric and Early Historic Settlement in the Kissim-
            mee River Valley: An Archaeological Survey of the Avon
            Park Air Force Range." *Florida Anthropologist* 40:287–300.

Barcia Carballido y Zúñiga, Andrés Gonzáles de
1951   *Chronological History of the Continent of Florida.* Translated

by Anthony Kerrigan. Gainesville: University of Florida Press.

Beriault, John, Robert Carr, Jerry Stipp, Richard Johnson, and Jack Meeder
1981   "The Archaeological Salvage of the Bay West Site, Collier County, Florida." *Florida Anthropologist* 34:39–58.

Biedma, Luys Hernández de
1904   "Relation of the Conquest of Florida." In *Narratives of the Career of Hernando de Soto,* edited by Edward Gaylord Bourne. New York: A. S. Barnes and Co.

Brain, Jeffrey P.
1985   "Introduction: Update of de Soto Studies Since the United States de Soto Expedition Commission Report." In *Final Report of the United States de Soto Expedition Commission,* by John R. Swanton. Washington, D.C.: Smithsonian Institution Press.

Brinton, Daniel G.
1859   *Notes on the Floridian Peninsula.* Philadelphia: Joseph Sabin.

Brown, Janice G., and Arthur D. Cohen
1985   "Palynologic and Petrographic Analyses of Peat Deposits, Little Salt Spring." *National Geographic Research* 1:21–31.

Brose, David S.
1979   "A Speculative Model of the Role of Exchange in the Prehistory of the Eastern Woodlands." In *Hopewell Archaeology, the Chillocothe Conference,* by Davis S. Brose and N'omi Greber. Kent, Ohio: Kent State University Press.

Bullen, Ripley P.
1957   "The Barnhill Mound, Palm Beach County, Florida." *Florida Anthropologist* 10:23–36.
1961   "Radiocarbon Dates for Southeastern Fiber-Tempered Pottery." *American Antiquity* 27:104–106.
1975   *A Guide to the Identification of Florida Projectile Points.* Gainesville: Kendall Books.
1978   "Tocobaga Indians and the Safety Harbor Culture." In *Tacachale: Essays on the Indians of Florida and Southeastern Georgia during the Historic Period,* edited by Jerald T. Milanich and Samuel Proctor. Gainesville: University of Florida Press.

Bullen, Ripley P., and Laurence E. Beilman
1973   "The Nalcrest Site, Lake Weohyakapka, Florida." *Florida Anthropologist* 26:1–22.

Bullen, Ripley P., and Harold K. Brooks
1967   "Two Ancient Florida Dugout Canoes." *Quarterly Journal of the Florida Academy of Sciences* 30:97–107.

Bullen, Ripley P., and Adelaide K. Bullen
1976   *The Palmer Site.* Florida Anthropological Society Publications 8.

Cabeza de Vaca, Alvar Nuñez
1984   "The Narrative of Alvar Nuñez Cabeza de Vaca." Edited by Frederick W. Hodge. In *Spanish Explorers in the Southern United States, 1528–1543,* edited by Frederick W. Hodge and Theodore H. Lewis. Austin: Texas State Historical Association.

Caldwell, Joseph R.
1964   "Interaction Spheres in Prehistory." In *Hopewellian Studies,* edited by Jospeh R. Caldwell and Robert L. Hall. Illinois State Museum Scientific Papers, Vol. 12.

Carr, Robert S.
1975   *An Archaeological and Historical Survey of Lake Okeechobee.* Florida Bureau of Historic Sites and Properties Miscellaneous Project Report Series 22.
1979   *An Archaeological and Historical Survey of the Site 14 Replacement Airport and Its Proposed Access Corridors, Dade County, Florida.* Washington, D.C.: Federal Aviation Administration.
1981   Salvage Excavation at Two Prehistoric Cemeteries in Dade County, Florida. Paper presented at 45th annual meeting, Florida Academy of Sciences, Winter Park.
1985   "Prehistoric Circular Earthworks in South Florida." *Florida Anthropologist* 38:288–301.
1986   "Preliminary Report on Excavations at the Cutler Fossil Site (8Da2001) in Southern Florida." *Florida Anthropologist* 39:231–232.
1990   "Archaeological Investigations at Pine Island, Broward County." *Florida Anthropologist* 43:249–261.

Carr, Robert S., and John G. Beriault
1984   "Prehistoric Man in South Florida." In *Environments of*

South Florida: Present and Past II, edited by Patrick J. Gleason. Miami Geological Society.

Carr, Robert S., M. Yasar Íscan, and Richard A. Johnson
1984   "A Late Archaic Cemetery in South Florida." *Florida Anthropologist* 37:172–188.

Chapman, Jefferson, and Gary D. Crites
1987   "Evidence for Early Maize (*Zea mays*) From the Icehouse Bottom Site, Tennessee." *American Antiquity* 52:352–353.

Chard, Chester S.
1975   *Man in Prehistory.* New York: McGraw-Hill.

Charlevoix, Pierre de
1966   *Journal of a Voyage to North-America.* Ann Arbor, Mich.: University Microfilms.

Chaves, Alonso de
1983   *Quatri Partitu en Cosmografia Practica, y por Otro Nombre Espejo de Navegantes.* Madrid: Instituto de Historia y Cultura Naval.

Clausen, Carl J., H. K. Brooks, and A. B. Wesolowsky
1975   *Florida Spring Confirmed as 10,000 Year Old Early Man Site.* Florida Anthropological Society Publication 7.

Clausen, Carl J., A. D. Cohen, Cesare Emiliana, J. A. Holman, and J. J. Stipp
1979   "Little Salt Springs, Florida: A Unique Underwater Site." *Science* 203:609–614.

Cockrell, Wilburn Allen
1970   Glades I and Pre-Glades Settlement and Subsistence Patterns on Marco Island (Collier County, Florida). Master's thesis, Florida State University, Tallahassee.

Cockrell, W. A., and Larry Murphy
1978   "Pleistocene Man in Florida." *Archaeology of Eastern North America* 6:1–13.

Coleman, Wesley F.
1989   "Salvage Excavations at the Trail Ridge Site, Dade County, Florida." *Florida Anthropologist* 42:257–262.

Connor, Jeannette Thurber
1925   *Colonial Records of Spanish Florida.* DeLand: Florida State Historical Society.

Covington, James W.
1959   "Trade Relations Between Southwestern Florida and Cuba—
          1680–1840." *Florida Historical Quarterly* 38:114–128.
1975   "Relations Between the Eastern Timucuan Indians and the
          French and Spanish, 1564–1567." In *Four Centuries of South-
          ern Indians,* edited by Charles M. Hudson. Athens: Univer-
          sity of Georgia Press.

Cumbaa, Stephen L.
1980   "Aboriginal Use of Marine Mammals in the Southeastern
          United States." *Southeastern Archaeological Conference Bul-
          letin* 17:6–10.

Curl, Donald W.
1986   *Palm Beach County.* Northridge, Calif.: Windsor Publica-
          tions.

Cushing, Frank Hamilton
1973   *Exploration of Ancient Key Dwellers' Remains on the Gulf
          Coast of Florida.* New York: AMS Press.

Daniel, I. Randolph, Jr., and Michael Wisenbaker
1987   *Harney Flats: A Florida Paleo-Indian Site.* Farmingdale, N.Y.:
          Baywood Publishing Co.

Davis, T. Frederick
1935   "Juan Ponce de León's Voyages to Florida." *Florida Histori-
          cal Quarterly* 14:1–70.

Denevan, William
1970   "Aboriginal Drained-Field Cultivation in the Americas."
          Science 169:647–654.

Díaz de Castillo, Bernal
1956   *The Discovery and Conquest of Mexico,* edited by Genaro
          Garcia, translated by A. P. Maudsley. New York: Farrar,
          Straus and Cudahy.

Dickinson, Jonathan
1981   *Jonathan Dickinson's Journal,* edited by Evangeline Walker
          Andrews and Charles McLean Andrews. Stuart: Florida Clas-
          sics Library.

Dobyns, Henry F.
1983   *Their Number Becomes Thinned.* Knoxville: University of
          Tennessee Press.

Doran, Glen H., and David N. Dickel
1988   "Raidometric Chronology of the Archaic Windover Ar-
            chaeological Site (8Br246)." *Florida Anthropologist* 41:365–
            380.

Douglass, A. E.
1885   "Ancient Canals on South-West Coast of Florida." *American
            Antiquarian and Oriental Journal* 7:277–285.
1890   "Description of a Gold Ornament from Florida." *American
            Antiquarian and Oriental Journal* 12:14–25.

Ehrenhard, John E.
1982   *Everglades National Park: Overview and Research Design.*
            Tallahassee: Southeast Archaeological Center, National
            Park Service.

Ehrenhard, John E., Robert S. Carr, and Robert C. Taylor
1978   *The Archeological Survey of Big Cypress National Preserve,
            Phase One.* Tallahassee: Southeast Archaeological Center,
            National Park Service.

Elvas, Gentleman of
1984   "The Narrative of the Expedition of Hernando de Soto, by
            the Gentlemen of Elvas." Edited by Theodore H. Lewis. In
            Spanish Explorers in the Southern United States, 1528–1543,
            edited by Frederick W. Hodge and Theodore H. Lewis.
            Austin: Texas State Historical Association.

Espenshade, Christopher Thomas
1983   Ceramic Ecology and Aboriginal Household Pottery Pro-
            duction at the Gauthier Site, Florida. Master's thesis, Uni-
            versity of Florida, Gainesville.

Fairbanks, Charles H.
1957   *Ethnohistorical Report of the Florida Indians.* Indian Claims
            Commission Dockets 73 and 151.
1968   "Florida Coin Beads." *Florida Anthropologist* 21:102–105.

Fairbridge, Rhodes W.
1984   "The Holocene Sea-Level Record in South Florida." In *Envi-
            ronments of South Florida: Present and Past II,* edited by
            Patrick J. Gleason. Miami Geological Society.

Farb, Peter
1968   *Man's Rise to Civilization as Shown by the Indians of North
            America from Primeval Times to the Coming of the Industrial
            State.* New York: E. P. Dutton.

120

BIBLIOGRAPHY

Felmley, Amy
1990   "Osteological Analysis of the Pine Island Site Human Remains." *Florida Anthropologist* 43:262–274.

Fernandez, Jose B.
1975   *Alvar Núñez Cabeza de Vaca, the Forgotten Chronicler.* Miami: Ediciones Universal.

Fontaneda, Do. d'Escalante
1973   *Memoir of Do. d'Escalante Fontaneda Respecting Florida.* Translated by Buckingham Smith, edited by David O. True. Miami: Historical Association of Southern Florida.

Ford, James A.
1969   *A Comparison of Formative Cultures in the Americas.* Smithsonian Contributions to Anthropology 11.

Fradkin, Arlene
n.d.   The Wightman Site: The Subsistence Pattern Among the Prehistoric Aborigines of Southwest Florida. Paper on file, Florida Museum of Natural History, Gainesville.
1976   The Wightman Site: A Study of Prehistoric Culture and Environment on Sanibel Island, Lee County, Florida. Master's thesis, University of Florida, Gainesville.

Furey, John F., Jr.
1972   The Spanish River Complex. Master's thesis, Florida Atlantic University, Boca Raton.

Gannon, Michael
1965   *The Cross in the Sand.* Gainesville: University of Florida Press.

Garcilaso de la Vega
1980   *The Florida of the Inca.* Translated by John Grier Varner and Jeannette Johnson Varner. Austin: University of Texas Press.

Gifford, John C., and Alfred H. Gilbert
1932   "Prehistoric Mounds in South Florida." *Science* 75:313.

Gilliland, Marion Spjut
1975   *The Material Culture of Key Marco, Florida.* Gainesville: University Presses of Florida.

Goggin, John W.
m.s.   The Archeology of the Glades Area, Southern Florida. M.s. in P. K. Yonge Library of Florida History, University of Florida, Gainesville.

121

1940   "The Tekesta Indians of Southern Florida." *Florida Histori-cal Quarterly* 18:274–284.

1944   "Archaeological Investigations on the Upper Florida Keys." *Tequesta* 4:13–35.

1947   "Manifestations of a South Florida Cult in Northwestern Florida." *American Antiquity* 13:273–276.

1950   "Stratigraphic Tests in the Everglades National Park." *American Antiquity* 3:228–246.

1964a   "Cultural Traditions in Florida Prehistory." In *Indian and Spanish Selected Writings*, edited by Charles H. Fairbanks, Irving Rouse, and William C. Sturtevant. Coral Gables: University of Miami Press.

1964b   "A Preliminary Definition of Archeological Areas and Periods in Florida." In *Indian and Spanish Selected Writings*, edited by Charles H. Fairbanks, Irving Rouse, and William C. Sturtevant. Coral Gables: University of Miami Press.

Goggin, John W., and William C. Sturtevant
1964   "The Calusa: A Stratified, Nonagricultural Society (With Notes on Sibling Marriage)." In *Explorations in Cultural Anthropology*, edited by Ward H. Goodenough. New York: McGraw-Hill.

Graves, Gypsy C.
1989   "Preliminary Report on the New River Midden (8Bd196) Excavation, Broward County, Florida." *Florida Anthropologist* 42:255–256.

Grayson, Donald K.
1977   "Pleistocene Avifauna and the Overkill Hypothesis." *Science* 185:691–692.

Griffin, John W.
m.s.   South Florida Archeology. M.s. for Handbook of North American Indians. Smithsonian Institution.

1943   "The Antillean Problem in Florida Archeology." *Florida Historical Quarterly* 22:86–91.

1952   "Prehistoric Florida: A Review." In *Archeology of Eastern United States*, edited by James B. Griffin. Chicago: University of Chicago Press.

1974   "Archeology and Environment in South Florida." In *Environments of South Florida: Present and Past*, edited by Patrick J. Gleason. Miami Geological Society Memoir 2.

1982   "Conclusions." In *Excavations at the Granada Site*, Vol. I, by

John W. Griffin et al. Tallahassee: Florida Division of Archives, History and Records Management.

1988   *The Archeology of Everglades National Park: A Synthesis.* Tallahassee: Southeast Archeological Center, National Park Service.

1989   "Time and Space in South Florida: A Synthesis." *Florida Anthropologist* 42:179–204.

Güemes y Horcasitas, Juan Francisco de

1991   "Governor Juan Francisco de Güemes y Horcasitas to the King, September 28, 1743." In *Missions to the Calusa,* edited and translated by John H. Hann. Gainesville: University of Florida Press.

Haggard, J. Villasana

1941   *Handbook for Translators of Spanish Historical Documents.* Archives Collection, University of Texas, Austin.

Hale, Stephen

m.s.   Evolution of Complex Forms of Social Organization in Prehistoric South Florida.

1984   "Prehistoric Environmental Exploitation Around Lake Okeechobee." *Southeastern Archaeology* 3:173–187.

Hammon, Briton

1760   *A Narrative of the Uncommon Sufferings, and Surprising Deliverance, of Briton Hammon.* Boston: Green and Russell.

Hann, John M., editor and translator

1991   *Missions to the Calusa.* Gainesville: University of Florida Press.

Hazeltine, Dan

1983   "A Late Paleo-Indian Site, Cape Haze Peninsula, Charlotte County, Florida." *Florida Anthropologist* 36:98–100.

Hewitt, George A., M.D.

1898   "Archeology of the West Coast of Florida." *The Medical Bulletin* 20:135–137, 173–174.

Higgs, Charles D.

1942   "Spanish Contacts with the Ais (Indian River) Country." *Florida Historical Quarterly* 21:25–39.

Hodge, Frederick Webb

1910   *Handbook of American Indians North of Mexico.* Bureau of American Ethnology Bulletin 30, Part One.

1911   *Handbook of American Indians North of Mexico*. Bureau of American Ethnology Bulletin 30, Part Two.

Hrdlicka, Ales

1918   "The 'Fossil' Man of Vero, Florida." In *Recent Discoveries Attributed to Early Man in Florida*. Bureau of American Ethnology Bulletin 66.

1922   *The Anthropology of Florida*. DeLand: Florida State Historical Society.

Johnson, William Gray

1990   "The Role of Maize in South Florida Aboriginal Native Societies: An Overview." *Florida Anthropologist* 43:209–214.

Jones, Calvin

1981   "Florida Anthropologist Interview with Calvin Jones, Part II, Excavations of an Archaic Cemetery in Cocoa Beach, Florida," conducted by Robert S. Carr. *Florida Anthropologist* 34:81–89.

Keegan, William F., ed.

1987   "Diffusion of Maize from South America: The Antillean Connection Reconstructed." In *Emergent Horticultural Economies of the Eastern Woodlands*. Southern Illinois University at Carbondale, Center for Archaeological Investigations, Occasional Paper 7.

Kennedy, W. Jerald, and M. Yasar Íscan

1987   "Archaeological Investigation of the Nebot Site (8PB219), Palm Beach, Florida." *Florida Scientist* 50:136–146.

Kenny, Michael

1970   *The Romance of the Floridas*. New York: AMS Press.

Kessel, Morton H.

1991   "The Role of Maize in South Florida Aboriginal Societies: A Comment." *Florida Anthropologist* 44:94.

Kottak, Conrad Phillip

1974   *Anthropology, the Exploration of Human Diversity*. New York: Random House.

Kroeber, A. L.

1953   *Cultural and Natural Areas of Native North America*. Berkeley: University of California Press.

Larson, Lewis H., Jr.

1980   *Aboriginal Subsistence Technology on the Southeastern*

Coastal Plain During the Late Prehistoric Period. Gainesville: University of Florida Press.

Lathrap, Donald W.
1987   "The Introduction of Maize in Prehistoric North America: The View from Amazonia and the Santa Elena Peninsula." In *Emergent Horticultural Economies of the Eastern Woodlands.* Southern Illinois University at Carbondale, Center for Archaeological Investigations, Occasional Paper 7.

Laudonnière, René
1975   *Three Voyages.* Translated by Charles E. Bennett. Gainesville: University Presses of Florida.

Laxson, D. D.
1957   "Three Small Dade County Sites." *Florida Anthropologist* 10:17-21.
1964   "Excavations in Southeast Florida, 1962-1963." *Florida Anthropologist* 17:177-181.
1966   "The Turner River Jungle Gardens Site." *Florida Anthropologist* 19:125-139.

Lazarus, William C.
1965   "Effects of Land Subsidence and Sea Level Changes on Elevation of Archaeological Sites on the Florida Gulf Coast." *Florida Anthropologist* 18:49-58.

Lewis, Clifford M.
1978   "The Calusa." In *Tacachale: Essays on the Indians of Florida and Southeastern Georgia during the Historic Period,* edited by Jerald T. Milanich and Samuel Proctor. Gainesville: University of Florida Press.

Long, Robert W.
1984   "Origin of the Vascular Flora of Southern Florida." In *Environments of South Florida: Present and Past II,* edited by Patrick J. Gleason. Miami Geological Society.

López, Feliciano
1991   "Fray Feliciano López to Fray Pedro Taybo, 1697." In *Missions to the Calusa,* edited and translated by John H. Hann. Gainesville: University of Florida Press.

López de Velasco, Juan
1991   "Memorial." In *Missions to the Calusa,* edited and translated by John H. Hann. Gainesville: University of Florida Press.

Lord, James S.
1989   "A Zoomorphic Bone Pine from Dade County, Florida."
        Florida Anthropologist 42:263-264.

Luer, George M.
1989   "Calusa Canals in Southwestern Florida: Routes of Tribute
        and Exchange." *Florida Anthropologist* 42:89-130.

Luer, George M., and Marion M. Almy
1982   "A Definition of the Manasota Culture." *Florida Anthropolo-
        gist* 35:34-58.

Luer, George M., et al.
1987   "The Myakkahatchee Site (8SO397), a Large Multi-Period
        Inland From the Coast Site in Sarasota County, Florida."
        Florida Anthropologist 40:137-153.

Lyon, Eugene
1967   *More Light on the Indians of the Ais Coast.* Term paper,
        University of Florida, Gainesville.
1976   *The Enterprise of Florida.* Gainesville: University Presses of
        Florida.

McGoun, William E.
1975   Some Observations on the State of Archaeology in Dade and
        Broward Counties. Term paper, Florida Atlantic Univer-
        sity, Boca Raton.
1981   Medals of Conquest in Calusa Florida. Master's thesis,
        Florida Atlantic University.
1985   The Everglades as an Ancient Highway. Paper presented at
        the 37th annual meeting, Florida Anthropological Society,
        Daytona Beach.
1986   "Sinkhole Supports Ice Age History." *The Palm Beach Post.*
        January 18:A28.
1987   Ancient Farmers Around the Everglades. Paper presented
        at the 39th annual meeting, Florida Anthropological Soci-
        ety, Clearwater.
1991   What Was That Strange Device? Paper presented at the 43rd
        annual meeting, Florida Anthropological Society, Pensa-
        cola.

McMichael, Alan Emerson
1982   A Cultural Resource Assessment of Horrs Island, Collier
        County, Florida. Master's thesis, University of Florida,
        Gainesville.

126

McMichael, Edward V.

1964   "Veracruz, the Crystal River Complex, and the Hopewellian Climax." In *Hopewellian Studies,* edited by Joseph R. Caldwell and Robert L. Hall. Illinois State Museum Scientific Papers Vol. 12.

Marquardt, William H.

1986   "The Development of Cultural Complexity in Southwest Florida: Elements of a Critique." *Southeastern Archaeology* 5 (1):63–70.

1987a   *Calusa News* No. 1. Institute of Archaeology and Paleo-environmental Studies, Florida Museum of Natural History, Gainesville.

1987b   South Floridian Contacts With the Bahamas: A Review and Some Speculations. Paper presented at the symposium, "Bahamas 1492: Its People and Environment," Freeport, Bahamas, November 1987.

1987c   The Calusa Social Formation in Protohistoric South Florida. In *Power Relations and State Formations,* edited by Thomas C. Patterson and Christine W. Gailey. Archeology Section, American Anthropological Association, Washington, D.C.

1988a   *Calusa News* No. 2. Institute of Archaeology and Paleo-environmental Studies, Florida Museum of Natural History, Gainesville.

1988b   Politics and Production among the Calusa of South Florida. In *Hunters and Gatherers,* volume 1: *History, Environment, and Social Change among Hunting and Gathering Societies,* edited by David Riches, Tim Ingold, and James Woodburn. Department of Anthropology, University College. London: Berg Publishers.

Masson, Marilyn A., et al.

1988   "The Taylor's Head Site (8BD74): Sampling a Prehistoric Midden on an Everglades Tree Island." *Florida Anthropologist* 41:336–350.

Matter, Robert Allen

1975   "Missions in Defense of Spanish Florida, 1566–1710." *Florida Historical Quarterly* 54:18–38.

Milanich, Jerald T.

1978a   "The Temporal Placement of Cushing's Key Marco Site, Florida." *American Anthropologist* 80:682.

1978b   "The Western Timucua: Patterns of Acculturation and.

127

Change." In *Tacachale: Essays on the Indians of Florida and Southeastern Georgia during the Historic Period*, edited by Jerald T. Milanich and Samuel Proctor. Gainesville: University of Florida Press.

1987   Corn and Calusa: DeSoto and Demography. In *Coasts, Plain, and Deserts: Essays in Honor of Reynold J. Ruppe*, edited by Sylvia W. Gaines. Tempe: Arizona State University, Anthropological Research Papers 38.

1989   "Where Did DeSoto Land? Identifying Bahia Honda." *Florida Anthropologist* 42:295–302.

Milanich, J. T., J. Chapman, A. S. Cordell, S. Hale, and R. A. Marrinan
1984   "Prehistoric Development of Calusa Society in Southwest Florida: Excavations on Useppa Island." In *Perspectives on Gulf Coast Prehistory*, edited by Dave D. Davis. Gainesville: University Presses of Florida.

Milanich, Jerald T., and Charles H. Fairbanks
1980   *Florida Archaeology*. New York: Academic Press.

Mitchem, Jeffrey McClain
1989   Redefining Safety Harbor: Late Prehistoric/Protohistoric Archaeology in West Peninsula Florida. Doctoral dissertation, University of Florida, Gainesville.

Moore, Clarence B.
1902   "Notes on the Ten Thousand Islands, Florida." *Journal of the Academy of Natural Sciences of Philadelphia* 13:458–470.

Mowers, Bert
1972   "Concretions Associated with Glades Prehistoric Sites." *Florida Anthropologist* 25:129–131.

1975   *Prehistoric Indian Pottery in South Florida*. Hollywood: Bert Mowers.

Neill, Wilfred T.
1955   "The Identity of Florida's 'Spanish Indians'." *Florida Anthropologist* 8:43–57.

Newman, Christine
1986   *Preliminary Report of Archaeological Investigations Conducted at the Cheetum Site, Dade County, Florida*. Miami: Archaeological and Historical Conservancy.

Newsom, Lee Ann, and Barbara Purdy
1990   "Florida Canoes: A Maritime Heritage from the Past." *Florida Anthropologist* 43:164–180.

Oviedo y Valdés, Gonzalo Fernández de
1904   "A Narrative of De Soto's Expedition Based on the Diary of Rogrigo Ranjel." In *Narratives of the Career of Hernando de Soto,* edited by Edward Gaylord Bourne. New York: A. S. Barnes and Co.

Palmer, Jay, and J. Raymond Williams
1977   "The Formation of Goethite and Calcareous Lenses in Shell Middens in Florida." *Florida Anthropologist* 30:24–27.

Payne, Claudine
1992   "Horr's Island Yields a New View of the Florida Archaic." *Calusa News* 6:1.

Purdy, Barbara
1981   *Florida's Prehistoric Stone Technology.* Gainesville: University Presses of Florida.

Reilly, Stephen Edward
1981   "A Marriage of Expedience: The Calusa Indians and Their Relations With Pedro Menéndez de Avilés in Southwest Florida." *Florida Historical Quarterly* 59:395–421.

Riley, Thomas J.
1987   "Ridged-Field Agriculture and the Mississippian Economic Pattern." In *Emergent Horticultural Economies of the Eastern Woodlands.* Southern Illinois University at Carbondale, Center for Archaeological Investigations, Occasional Paper 7.

Romans, Bernard
1940   *A Concise Natural History of East and West Florida.* Jacksonville: Federal Writers Program.

Romero, Francisco
1991   "Statement by the Ensign Francisco Romero." In *Missions to the Calusa,* edited and translated by John H. Hann. Gainesville: University of Florida Press.

Rouse, Irving
1940   "Some Evidence Concerning the Origins of West Indian Pottery-Making." *American Anthropologist* (New Series) 42:49–80.
1951   *A Survey of Indian River Archeology, Florida.* Yale University Publications in Anthropology 44.

Royal, William, and Eugenie Clark
1960   "Natural Preservation of Human Brain, Warm Mineral Springs, Florida." *American Antiquity* 26:285–287.

Russo, Michael
1985   *Zaremba: A Short-Term Use Malabar II Site.* Department of Anthropology Miscellaneous Reports Series No. 25, Florida State Museum, Gainesville.

Safford, W. E.
1919   "Natural History of Paradise Key and the Nearby Everglades of Florida." *Annual Report of the Board of Regents of Smithsonian Institution* for 1916–1917.

Ste. Claire, Dana
1990   "The Archaic in East Florida: Archaeological Evidence from Early Coastal Adaptations." *Florida Anthropologist* 43:189–197.

Saunders, Lorraine P.
1972   Osteology of the Republic Groves Site. Master's thesis, Florida Atlantic University, Boca Raton.

Scarry, C. Margaret
1982   "Paleothnobotany of the Granada Site." In *Excavations at the Granada Site.* Vol. I, John W. Griffin et al. Tallahassee: Florida Division of Archives, History and Records Management.

Schell, Rolfe E.
1966   *De Soto Didn't Land at Tampa.* Fort Myers Beach: Island Press.

Scholl, David W., Frank C. Craighead, Sr., and Minze Stuiver
1969   "Florida Submergence Curve Revisited: Its Relation to Coastal Sedimentation Rates." *Science* 163:562–564.

Schwehm, Alice Gates
1983   The Carved Wood Effigies of Fort Center, a Glimpse of South Florida's Prehistoric Art. Master's thesis, University of Florida, Gainesville.

Sears, William H.
m.s.   Southeastern United States—400 B.C.–1000 A.D. m.s. for Handbook of North American Indians. Washington, D.C.: Smithsonian Institution.
1958   "Burial Mounds on the Gulf Coastal Plain." *American Antiquity* 23:274–284.

130

1971   "Food Production and Village Life in Prehistoric Southeast-
        ern United States." *Archaeology* 24:322–329.
1977   "Seaborne Contacts Between Early Cultures in Lower South-
        eastern United States and Middle Through South America."
        In *The Sea in the Pre-Columbian World,* edited by Elizabeth
        P. Benson. Washington, D.C.: Trustees for Harvard Univer-
        sity.
1982   *Fort Center: An Archaeological Site in the Lake Okeechobee
        Basin.* Gainesville: University Presses of Florida.

Sellards, E. H.
1916   *Human Remains and Associated Fossils from the Pleistocene
        of Florida.* Reprinted from Eighth Annual Report of the
        Florida State Geological Survey, 121–160, Plates 15–31.

Small, John Kunkel
1929   *From Eden to Sahara.* Lancaster, Pa.: Science Press Printing
        Co.

Smith, Bruce D.
1986   "The Archaeology of the Southeastern United States: From
        Dalton to de Soto, 10,500–500 B.P." In *Advances in World
        Archaeology,* Vol. 5, edited by Fred Wendorf and Angela E.
        Close. New York: Academic Press.
1989   "Origins of Agriculture in Eastern North America." *Science*
        246:1566–1571.

Smith, Hale G.
1951   "The Ethnological and Archeological Significance of
        Zamia." *American Anthropologist* 53:238–244.

Smith, Hale G., and Mark Gottlob
1978   "Spanish-Indian Relationships: Synoptic History and Ar-
        chaeological Evidence, 1500–1763." In *Tacachale: Essays on
        the Indians of Florida and Southeastern Georgia during the
        Historic Period,* edited by Jerald T. Milanich and Samuel
        Proctor. Gainesville: University of Florida Press.

Solís de Merás, Gonzalo
1964   *Pedro Menéndez de Avilés.* Translated by Jeannette Thurber
        Conner. Gainesville: University of Florida Press.

Spencer, Robert F., and Jesse D. Jennings
1965   *The Native Americans.* New York: Harper and Row.

Stahl, Jeremy D.
1986   An Ethnohistory of South Florida, 1500–1575. Master's thesis, University of Florida, Gainesville.

Steele, William D.
1972   *The Wilderness Tattoo.* New York: Harcourt Brace Jovanovich.

Stewart, T. D.
1946   *A Reexamination of the Fossil Human Skeletal Remains from Melbourne, Florida, with Further Data on the Vero Skull.* Smithsonian Miscellaneous Collections, Vol. 106, No. 10.

Stirling, M. W.
1931   "Mounds of the Vanished Calusa Indians of Florida." In *Explorations and Field-Work of the Smithsonian Institution* in 1930.
1935   "Smithsonian Archeological Projects Conducted Under the Federal Emergency Relief Administration, 1933–1934." In *Smithsonian Report* for 1934.
1936   "Florida Cultural Affiliations in Relation to Adjacent Areas." In *Essays in Anthropology.* Freeport, N.Y.: Books for Libraries Press.
1940   "The Historic Method as Applied to Southeastern Archeology." In *Essays in Historical Anthropology of North America.* Smithsonian Miscellaneous Collections 100.

Stone, Doris
1939   "The Relationship of Florida Archaeology to That of Middle America." *Florida Historical Quarterly* 17:211–218.

Struever, Stuart
1972   "The Hopewell Interaction Sphere in Riverine-Western Great Lakes Culture History." In *Contemporary Archaeology,* edited by Mark P. Leone. Carbondale: Southern Illinois University Press.

Sturtevant, William C.
1953   "Chakaika and the 'Spanish Indians'." *Tequesta* 13:35–74.
1960   *The Significance of Ethnological Similarities Between Southeastern North America and the Antilles.* Yale University Publications in Anthropology 64.
1962   "Spanish-Indian Relations in Southeastern North America." *Ethnohistory* 9:41–94.
1978   "The Last of the South Florida Aborigines." In *Tacachale: Essays on the Indians of Florida and Southeastern Georgia*

during the Historic Period, edited by Jerald T. Milanich and Samuel Proctor. Gainesville: University of Florida Press.

Swanton, John R.

1922   *Early History of the Creek Indians and Their Neighbors.* Bureau of American Ethnology Bulletin 73.

1979   *The Indians of the Southeastern United States.* Washington, D.C.: Smithsonian Institution Press.

1985   *Final Report of the United States DeSoto Expedition Commission.* Washington, D.C.: Smithsonian Institution Press.

Tesar, Louis D.

1989   "The Case for Concluding that De Soto Landed near Present-Day Fort Myers, Florida: The Conclusions Presented by Warren H. Wilkinson Reviewed." *Florida Anthropologist* 42:276–279.

Valdés, Gerónimo

1991   "Bishop Gerónimo Valdés to the King, December 9, 1711." In *Missions to the Calusa,* edited and translated by John H. Hann. Gainesville: University of Florida Press.

Van Beck, John C., and Linda M. Van Beck

1965   "The Marco Midden, Marco Island, Florida." *Florida Anthropologist* 18:1–20.

Vargas Ugarte, Ruben

1935   "The First Jesuit Mission in Florida." *Historical Records and Studies, United States Catholic Historical Society* 25:59–148.

Walker, Karen Jo

1992   "The Mystery of the Pineland Canal." *Calusa* 6:3.

Waring, A. J., and Preston Holder

1968   "A Prehistoric Ceremonial Complex in the Southeastern United States." In *The Waring Papers.* Cambridge, Mass.: Papers of the Peabody Museum, Harvard University.

Weed, Carol S., L. Janice Campbell, and Prentice M. Thomas

1982   *Literature Review and Cultural Resources Survey of the U.S. Coast Guard Light Station, Jupiter Inlet, Palm Beach County, Florida.* New World Research Inc. Report of Investigations No. 59.

Wenhold, Lucy L.

1936   *A 17th Century Letter of Gabriel Diaz Vara Calderón, Bishop of Cuba, Describing the Indians and Indian Missions of Florida.* Smithsonian Miscellaneous Collections 95, No. 16.

Wharton, Barry R., George R. Ballo, and Mitchell E. Hope
1981 "The Republic Groves Site, Hardee County, Florida." *Florida Anthropologist* 34:59–80.

Wheeler, Ryan J.
1992 "The Riviera Complex: An East Okeechobee Archaeological Area Settlement." *Florida Anthropologist* 45:5–17.

Widmer, Randolph J.
1974 *A Survey and Assessment of Archaeological Resources on Marco Island, Collier County, Florida.* Florida Bureau of Historic Sites and Properties, Miscellaneous Report Series No. 18.
1988 *The Evolution of the Calusa: A Nonagricultural Chiefdom on the Southwest Florida Coast.* Tuscaloosa: University of Alabama Press.
1989 "The Relationship of Ceremonial Artifacts from South Florida with the Southeastern Ceremonial Complex." In *The Southeastern Ceremonial Complex: Artifacts and Analysis,* edited by Patricia Galloway. Lincoln: University of Nebraska Press.

Wilkinson, Warren H.
1947 "The DeSoto Expedition in Florida." Estero, Florida: *The American Eagle.* [Various issues beginning November 6.]

Will, Lawrence E.
1977 *Cracker History of Okeechobee.* Belle Glade: Glades Historical Society.

Willey, Gordon R.
1949 *Excavations in Southeast Florida.* Yale University Publications in Anthropology 42.
1982 *Archeology of the Florida Gulf Coast.* Smithsonian Miscellaneous Collections 113. Florida Book Store Reprint Edition.

Willey, Gordon R., and Philip Phillips
1958 *Method and Theory in American Archaeology.* Chicago: University of Chicago Press.

Williams, John Lee
1962 *The Territory of Florida.* Gainesville: University of Florida Press.

Williams, Lindsey
1989 "A Charlotte Harbor Perspective on De Soto's Landing Site." *Florida Anthropologist* 42:280–294.

134

Wing, Elizabeth S.
1965   "Animal Bones Associated with Two Indian Sites on Key
          Marco." *Florida Anthropologist* 18:21–28.
1977   "Factors Influencing Exploitation of Marine Resources." In
          The Sea in the Pre-Columbian World, edited by Elizabeth P.
          Benson. Washington, D.C.: Trustees for Harvard Univer-
          sity.
1978   "Subsistence at the McLarty Site, Indian River County."
          Florida Anthropologist 31:3–7.

Wright, J. Leitch, Jr.
1981   *The Only Land They Knew.* New York: Free Press.

Wymer, Dee Anne
1987   "The Middle Woodland-Late Woodland Interface in Central
          Ohio: Subsistence Continuity and Cultural Change." In
          Emergent Horticultural Economies of the Eastern Woodlands.
          Southern Illinois University at Carbondale, Center for Ar-
          chaeological Investigations, Occasional Papers 7.

Zubillaga, Felix
1946   *Monumenta Antiquae Floridae.* Rome: Jesuit Historical Insti-
          tute.
1991   "Report on the Florida Missions by Father Juan Rogel,
          Written Between the Years 1607 and 1611." In *Missions to the
          Calusa,* edited and translated by John H. Hann. Gainesville:
          University of Florida Press.

Index

INDEX

ABOUT THE AUTHOR

William E. McGoun is a professional journalist of thirty-four years and currently serves as senior editorial writer of *The Palm Beach Post* in West Palm Beach, Florida. He began his anthropological studies on a part-time basis in 1974, earning a bachelor's degree in 1977 and a master's degree in 1981, both from Florida Atlantic University, Boca Raton. He took his doctoral class work at the University of Florida during the 1981–1982 school year, as a Graduate Council Fellow.

Besides numerous newspaper articles, he has had articles published in *Florida Journal of Anthropology, Journal of Cherokee Studies,* and *The Reflector.* He is the author of two published history books, *A Biographical History of Broward County* (1972) and *Hallandale* (1976).